Applications of Data Science to Molecules and Materials 2019

AA002985

Topical Conference at the 2019 AIChE Annual Meeting

Orlando, Florida, USA
10-15 November 2019

ISBN: 978-1-7138-0553-3

Printed from e-media with permission by:

Curran Associates, Inc.
57 Morehouse Lane
Red Hook, NY 12571

Some format issues inherent in the e-media version may also appear in this print version.

Copyright© (2019) by AIChE
All rights reserved.

Printed with permission by Curran Associates, Inc. (2020)

For permission requests, please contact AIChE
at the address below.

AIChE
120 Wall Street, FL 23
New York, NY 10005-4020

Phone: (800) 242-4363
Fax: (203) 775-5177

www.aiche.org

Additional copies of this publication are available from:

Curran Associates, Inc.
57 Morehouse Lane
Red Hook, NY 12571 USA
Phone: 845-758-0400
Fax: 845-758-2633
Email: curran@proceedings.com
Web: www.proceedings.com

TABLE OF CONTENTS

556963 Deep Learning of Activation Energies and Automated Reaction Dataset Generation

Automated mechanism generation is an efficient tool for rapidly constructing large-scale chemical reaction mechanisms, which are necessary for realistic modeling of complex processes, such as combustion and oxidation, soot formation, or pyrolysis. Nonetheless, hard-coded reaction templates employed by mechanism generation software do not cover the set of all possible reactions that might be important. Moreover, only a limited quantity of high-quality data is available for the estimation of kinetic rates during mechanism generation, which may result in poor estimates and the erroneous exclusion of important reactions. In response, we have developed methods to automatically generate a large database of chemical reactions and designed a template-free deep learning model capable of providing a quick assessment of potentially important reactions that are missing from a reaction mechanism.

We use activation energy as a temperature-independent measure to gauge the potential importance of new reactions and train a machine learning model to estimate the energies by relying on a reaction encoding which predominantly focuses on the atoms involved in the reactive center. In order to predict kinetically relevant activation energies as well as being able to tell which reactions have barriers that are so large that they cannot possibly be relevant, the training data for the machine learning model must span a more diverse set of reactions than is currently available in the literature. We use an automated transition state finding algorithm to generate a diverse training set of tens of thousands of reactions, and thus create an activation energy prediction model that can quantify the importance of proposed reactions.

Copyright © American Institute of Chemical Engineers. All rights reserved.

566431 Towards Automated Discovery of Plausible Reaction Paths in Complex Catalytic Systems Using Network Generation and Optimization

The application of state-of-the-art computational tools, viz. density functional theory (DFT) to calculate the energetics of every species and transition state and microkinetic modeling to obtain reaction rates and yields is computationally intractable for complex reaction networks comprising of thousands of reactions and intermediates. In this talk, we present a computationally tractable method to identify dominant reaction pathways for such complex heterogenous catalytic reaction systems. The method: (i) constructs the reaction network in an automated manner using the knowledge of potential elementary steps, (ii) uses simple data-driven models derived from DFT calculations that can then be applied to the rest of the network, (iii) employs a mixed-integer optimization problem to "screen" for the dominant pathways based on energetic feasibility metrics or maximum reaction rate estimates, (iv) rank orders potential alternative solutions of the optimal pathways within the errors of density functional theory and the data-driven model, and (v) generates volcano plots based on known linear scaling correlations for catalyst screening.

The details of the proposed method will be discussed in the context of demonstrative examples comprising of oxygenate decomposition and dehydrogenation of organic hydrogen carriers on transition metal catalysts.

Copyright © American Institute of Chemical Engineers. All rights reserved.

572387 Robust Catalyst Characterization and Estimation of Mechanistic Pathways through Data Science

Robust Catalyst Characterization and Estimation of Mechanistic Pathways through Data Science

Transient kinetic experiments provide a significant volume of data as the material is evaluated as a function of time rather than at a singular steady-state. This allows more detailed microkinetic information to be collected where variations in temperature, pressure or concentration on the materials surface is reflected in the measured response. However, the resolution of this detail also requires the development of new analysis to accurately inform the user of the underlying mechanism while maintaining the temporal nature of the data. A novel approach presented here utilizes a combination of modern machine learning techniques and physics-based assumptions to analyze the full information available from transient kinetic experiments without assumption of a kinetic model. Furthermore, this approach applies feature selection within a global reaction network to provide an optimal trade-off between the selection of a complex estimated mechanism and the minimization of the fit on the measured reaction rate. As an example, a set of catalysts are compared when performing ammonia decomposition in transient pulse response experiments. We were able to quantitatively measure the ammonia gas interaction with different surface species and the degree of impact on the transformation rate. As such, the global sparse covariance matrix describes the network path of individual catalysts. More specifically, we are able to distinguish differences in each catalyst based on the degree, if at all, of the complete set of surface and gas phase interaction combinations. The figure below shows an example of the obtained network where the surface species are indicated as *z* and the gas species are indicated as *c*. The coloration of each node indicates the grouping of the interactions while the edges indicate a significant interaction. These connections between the gas and surface species are remarkably distinct for different materials and represent a mechanistic signature for discriminating catalysts. The results of this novel method provide interpretability through integration of physical assumptions, estimation of the mechanistic reaction pathways and a unique experimental kinetic fingerprint for each material whereby catalysts can be microkinetically discriminated.

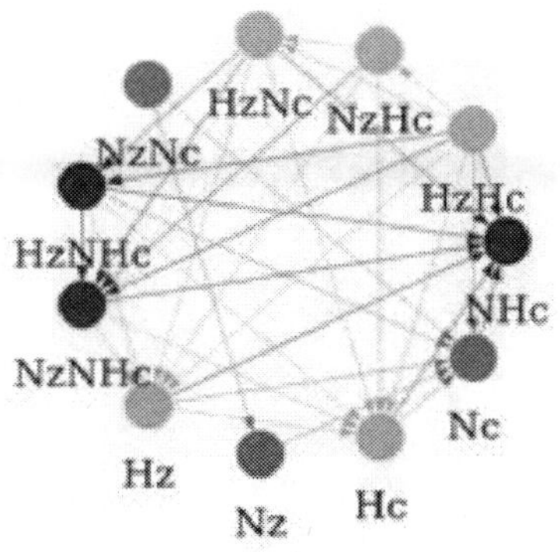

Copyright © American Institute of Chemical Engineers. All rights reserved.

573166 Autoneb: A Toolkit for Automating Reaction Network Calculation

We present an automated toolkit for the first-principles studies of reaction pathways, especially in complicated reaction networks involving multiple conformers of the reaction species. Challenges remain in the automatic generation of conformers of an arbitrary molecule due to the high variability in bonding and charge state and the limited availability of suitable force field parameters needed in conventional conformer search methods, especially for charged species or radicals. We overcome this through automated enumeration of all possible combination of rotatable dihedrals, followed by automated pruning to generate featured conformers. Pruning of the conformers is achieved through a 3-step process: (1) elimination of conformers with atom clashes based on molecular fingerprint, (2) geometry optimization with density functional theory and elimination of energetically unfavorable conformers, (3) extraction of featured conformers via a clustering algorithm based on the geometries. For the surviving conformers, nudged elastic band (NEB) calculations are then automatically set up for each possible reactant-product conformer pair in the reaction network. A new interpolation scheme is implemented in AutoNEB for generating the initial structures of the "beads" in NEB, in order to properly handle reactions involving complicated motion in the internal coordinates. Finally, automated post-processing aids the discovery of the most favorable route in the whole network, with visualization of the energy profile. This toolkit has been applied to the reaction network for gas-phase Pomeranz-Fritsch synthesis of isoquinoline and has provided an insightful explanation of the phenomena observed in mass spectroscopy.

Copyright © American Institute of Chemical Engineers. All rights reserved.

575887 Recent Advances in Automated Transition State Theory Calculations: A Case Study Using Butanol

With the effects of global climate change becoming more apparent, researchers have been pushing for a shift from conventional fossil fuels to greener alternatives. However, these fuels need to be adequately studied *a priori* to make reasonable experimental choices. In these studies, detailed reaction mechanisms need to be constructed to explain the complex combustion phenomena. Because of the sheer number of thermodynamic and kinetic parameters, they are often estimated rather than determined experimentally. *Ab initio* calculations and transition state theory (TST) can determine thermodynamic and kinetic parameters more accurately than commonly used estimates. However, given the number of reactions and species present in even simple combustion systems, these calculations need to be automated.

Many researchers have been working on tools to perform these automated TST calculations, but for this work we aim to improve AutoTST, the automated transition state theory code first developed by Bhoorasingh et al. [1]. The first generation of AutoTST would start by creating a transition state geometry guess by using a functional group tree to decide on key distances in the reaction center and using RDKit [2], an open source cheminformatics toolkit, to generate the starting geometry. AutoTST would then employ Gaussian [3] to perform a series of partial optimizations to arrive at an optimized transition state complex. This complex is then validated through automated intrinsic reaction coordinate (IRC) calculations and validated geometries are then fed into Arkane [4] to obtain kinetic parameters. This work has improved AutoTST by: rewriting AutoTST as an independent, reusable, module; including the effects of 1-D hindered rotors; adding a systematic conformer analysis; and bypassing the expensive IRC calculations by using vibrational analysis to validate transition states.

To assess the impact of these improvements, we performed automated TST calculations on 1117 reactions and 298 species present in a model for the combustion of butanol [6] and compared these reaction rates and thermodynamic expressions against values calculated by the first generation of AutoTST, values estimated by RMG, and values published in the model. From here, we identified reactions and species where our calculations deviated the most from comparisons, perform benchmark calculations to generate high fidelity kinetics and thermodynamic parameters for these reactions and species, and compare and investigate the sources of these discrepancies. This framework will enable researchers the ability to generate large amounts of high fidelity kinetic and thermodynamic training data on-the-fly for machine-learning based estimation methods.

References:

[1] Bhoorasingh, P. L.; Slakman, B. L.; Khanshan, F. S.; Cain, J. Y.; West, R. H. Automated Transition State Theory Calculations for High-Throughput Kinetics. J. of Phys. Chem. A. 2017, 121, 6896-6904.

[2] RDKit: Open-source cheminformatics; http://www.rdkit.org

[3] Gaussian 16, Revision B.01, Frisch M. J. et al., Gaussian, Inc., Wallingford CT, 2016.

[4] Allen, J. W.; Green, W. H.; Arkane: http://reactionmechanismgenerator.github.io/RMG-Py/users/arkane/credits.html

[5] Gao, C. W.; Allen, J. W.; Green, W. H.; West, R. H. Reaction Mechanism Generator: Automatic Construction of Chemical Kinetic Mechanisms. Comput. Phys. Commun. 2016, 203, 212–225.

[6] Sarathy S. M. et al. A comprehensive chemical kinetic combustion model for the four butanol isomers. Comb. Flame. 2016, 159, 2028-2055.

Copyright © American Institute of Chemical Engineers. All rights reserved.

577241 Smart Synthetic Path Generation System with Analysis of Reaction Risk Enhanced By Automated Open Data Collection

Chemical accidents often occur during laboratory experiments, pilot plant operations and abnormal reactor operations. To prevent these incidents, it is necessary to find relevant information before starting the synthesis experiment as well as process operations. In the process design stage, reaction information is essential to prevent reaction runaway. To get information, there are various sources of information related to synthesis, including the Internet and literature, and it takes a long time to search. In order to solve this problem, we propose an intelligent support system for synthetic path generation.

The DB used in the system is constructed by collecting open data about chemicals, synthesis, and safety (e.g., NFPA 704, GHS, MSDS, PAC, etc.). The data about chemical substances are available in freely accessible chemical DBs, which can be found in PubChem, ChEMBL, etc., and synthesis data can also be found through various sources. However, it is difficult to organize and search through this various set of data. For this reason, Web Scraping and Web Crawling techniques based on Python are used to explore, collect, organize, and obtain data from diverse sources. Since safety information offered by each Web site is different, safety data for chemicals are collected separately and made into a unified DB. The path generation algorithm is a combination of Depth-First Search and pruning algorithm. The suggested algorithm proceeds with the search based on the target substance and cuts off the child nodes if it is judged that there is no possibility for the child nodes. As a result, only the nodes that have possibility of synthesis are left, and paths are generated by connecting the nodes.

The results of generation show a synthetic path graph of various steps and information of each path for the target substance. This graph shows possible synthetic paths connecting from the target substance node to the starting material node and the user can easily assess the whole path. Synthesis data nodes containing hazardous chemicals are graded and displayed in a separate color (green, yellow or red depending on the severity). Each path includes data such as operating conditions and chemicals used for each synthesis method on the path. For each path where hazardous chemicals are used, additional information is displayed by analyzing the risk in the aspect of safety, health, and environment. After analyzing the synthesis data, the risk class of each synthetic path is classified and weighted for ranking.

The data in the built-in DB are continuously collected and automatically updated. First, about 100,000 synthesis data were obtained and the system test was conducted. We used Scifinder and SciPlanner of CAS to verify the data obtained from the test. Scifinder has a large DB based on many literatures and patents, so it is easy to search for synthetic information. As a result of using SciPlanner, there are few cases where there is no result corresponding to the proposed system's results, but most of matching results could be found. It verifies that actually available paths are all generated using the system even if the size of DB becomes large. The system is being released as open source for free use. For the benefit of each research institution, researchers can register their private data and expand the DB according to the format type. It increases the likelihood of getting results that are tailored to their goals. Using the proposed system, it is expected that researchers will find a safer, reliable, and economical reaction pathway by referring to the generated results and help prevent accidents in experiments and eventual plant operations.

Copyright © American Institute of Chemical Engineers. All rights reserved.

572321 Reaction Network Analysis and Motifs for Rhythmic Dynamics in Metabolic and Circadian Models of Cyanobacteria in a Photobioreactor

Reaction network theories are tools for stability analysis of open reacting systems provided that stoichiometric (chemical) equations are given for each reaction step together with power law rate expressions. Based on stoichiometry alone, elementary subnetworks (known also as elementary modes or extreme currents) are identified and their capacity for displaying dynamical instabilities, such as bistability and oscillations, is evaluated by examining the associated Jacobian matrix [1]. This analysis is qualitative in the sense that only reaction orders are needed as input information, whereas rate coefficients and concentrations of chemical components may remain unspecified. In the next step, the subnetworks are combined to form the entire network and its stability is deduced from the stability of the constituting subnetworks. This combination principle can be conveniently used for kinetic parameter estimation of the unknown/unspecified rate coefficients by applying linear optimization to a set of constraint equations balancing linearly combined subnetworks with the corresponding rate expressions [2]. From mathematical point of view, this is a special case of convex optimization.

From the application point of view, the outlined theory is applied to experimentally measured activity of photosynthesizing diazotrophic cyanobacteria in a photobioreactor [3]. In particular, we focus on biochemical processes giving rise to experimentally observed change from a steady state to oscillatory dynamics. The oscillations in cyanobacteria come from either circadian cycles synchronized with external light/dark cycle or from an internal ultradian clock [4], which is active even in the absence of external environmental cues. For the former, we examine models of circadian clock associated with a network involving the KaiABC proteins and their regeneration via a transcriptional network, for the latter, a carbon-nitrogen metabolic model is analyzed.

For these biochemical systems a reaction mechanism is assumed (or available from previous research) with only a limited set of known kinetic parameters, in addition to input/output parameters known from the experiment. Then the set of unknown kinetic parameters is estimated via linear optimization so that the dynamics displayed by the model coincides with the experimentally observed behavior (emergence of oscillations). Such an estimate may not yield the ultimate best fit, rather, it helps to locate a region in the parameter space, where the observed dynamics are reproduced by the model. This global approach is useful especially when the number of unknown parameters is large. Once a suitable parameter region is found, either standard least-square methods or other more refined algorithms involving deep learning may be used to fine tune the parameters values.

In addition, the reaction network theory is useful in identifying subnetworks that are at the core of destabilizing the steady state. In particular, such subnetworks provide prototypes/motifs of (bio)chemical oscillators, which possess a characteristic network topology [5]. Thus the search for dynamical instabilities in large networks typical for biosystems can be reduced to the search for motifs.

References:

[1] B. L. Clarke, Stability of complex reaction networks. Advances in Chemical Physics 43, 1-278 (1980).

[2] V. Radojkovic and I. Schreiber, Constrained Stoichiometric Network Analysis, Physical Chemistry Chemical Physics 20, 9910–9921 (2018).

[3] J. Červený, J. Šalagovič, F. Muzika, D. Šafránek and I. Schreiber, Influence of Circadian Clocks on Optimal Regime of Central C-N Metabolism of Cyanobacteria. In: Cyanobacteria: From Basic Science to Applications, Edited by A.K. Mishra, D.N. Tiwari and A.N. Rai, Elsevier, London, 2019, pp. 193-206.

[4] J. Červený, M. A. Sinetova, L. Valledor, L. A. Sherman, L. Nedbal, Ultradian metabolic rhythm in the diazotrophic cyanobacterium Cyanothece sp. ATCC 51142. Proc. Natl. Acad. Sci. 110 (32), 13210–13215

(2013).

[5] J. Ross, I. Schreiber and M. O. Vlad, Determination of Complex Reaction Mechanisms, Oxford University Press, Inc., New York, 2006.

Copyright © American Institute of Chemical Engineers. All rights reserved.

579605 Data Science for Assembly Engineering

Discovery and design of new materials able to self assemble from nanoscale building blocks are becoming increasingly enabled by large-scale molecular simulation. Aided by fast simulation codes leveraging powerful computer architectures, an unprecedented amount of data can be generated in the blink of an eye, shifting the effort and focus of the computational scientist from the simulation to the data. How do we manage so much data, and what do we do with it when we have it? In this talk, we discuss the applications of data science and data-driven thinking to molecular and materials simulation. Although we do so in the context of assembly engineering of soft matter, the tools and techniques discussed are general and applicable to a wide range of problems. We present applications of machine learning to automated, structure identification of complex colloidal crystals, high-throughput mapping of phase diagrams, the study of kinetic pathways between fluid and solid phases, and the discovery of previously elusive design rules and structure-property relationships.

We also discuss new tools we've developed to support these studies. We present *pythia*, a wrapper library for use with data analysis tools that generates descriptors of particle systems for ML. We present *signac*, an open-source data management framework that facilitates the management of large and heterogeneous data spaces by creating a well-defined, indexable storage layout for data and metadata. *signac* is lightweight and manages computational workflows on platforms ranging from laptops to supercomputers. *signac* supports HDF5 integration for rapid access to large numerical data arrays, has new tools to import and export data spaces for long-term archival and publication of data sets; and integrates seamlessly with the scientific Python ecosystem for use with Jupyter notebooks and more. *signac*is application agnostic and has been used for molecular simulations, quantum chemistry, photonics, computational fluid dynamics, machine learning, graph mining, and even organizing experimental data.

Contributors include Carl Simon Adorf, Joshua A. Anderson, Chengyu Dai, Bradley Dice, Yina Geng, Vyas Ramasubramani, and Matthew Spellings.

Copyright © American Institute of Chemical Engineers. All rights reserved.

580104 A Computational Roadmap for Better Electronic Noses

Most gas sensors can only detect a single gas species (e.g., a household carbon monoxide detector), with a small subset able to distinguish among a few gas species (e.g., commercial "electronic nose" sensors). However, few attempts have been made – and none successfully – at creating a device that could detect a very broad spectrum of gases under a wide range of operating conditions. This seemingly challenging task is carried out every day by our own biological noses, however. Creating an artificial nose as capable as a biological nose would have enormous socioeconomic impact; analogous to the invention of the camera (i.e., the "artificial eye"). It would be transformative for a wide range of industrial and military applications, but, in particular, it would revolutionize health diagnostics.

So what would it take to build a better electronic nose? Electronic noses rely on arrays of chemically distinct materials that each interact uniquely with different gas species, which is closely analogous to how biological noses work. Current electronic noses, however, use small arrays with typically less than 20-30 materials. Dogs, on the other hand, have thousands of distinct olfactory receptors, each of which are replicated tens of thousands of times throughout the interior of their noses. Recent evidence suggests that building larger arrays can significantly improve electronic nose performance, but only if the sensing materials are carefully chosen. For large arrays this becomes a classic "big data" problem, requiring significant computational power and clever algorithms to address the combinatorial explosion of possible arrays one could construct.

In this talk, we discuss recent work on the computational design of gas sensing arrays, based on modeling gas adsorption in metal-organic frameworks (MOFs), and lay out a roadmap for how we might eventually build an electronic nose that could rival a dog's one.

Copyright © American Institute of Chemical Engineers. All rights reserved.

579610 Using Artificial Intelligence to Discover New Materials: The Role of Material Representations

Rational, data-driven materials discovery has the potential to make research and development efforts far faster and cheaper. In such a paradigm, computer models trained to find patterns in massive chemical datasets would rapidly scan compositions and systematically identify attractive candidates. Here, we present several examples of our work on developing machine learning (ML) and deep learning methods capable of creating predictive models using a diverse range of materials data. We illustrate three distinct materials representations and compare the resulting ML models: 1) composition-only attributes, 2) crystal structure-based attributes based on Voronoi tesselation, and 3) an improved version of the crystal graph convolutional neural net (iCGCNN). We also examine the use of "representation-free" approaches based on neural net deep learning. As input training data, we demonstrate ML on both large computational datasets of DFT calculations, as implemented in the Open Quantum Materials Database (oqmd.org), and also experimental databases of materials properties. We construct ML models using a large and chemically diverse list of attributes, which we demonstrate can be used as an effective tool to automatically learn intuitive design rules, predict diverse properties of crystalline and amorphous materials, such as formation energy, specific volume, band gap energy, and glass-forming ability, and accelerate combinatorial searches.

Copyright © American Institute of Chemical Engineers. All rights reserved.

578305 CAR-Mediated T Cell Activation: Linking Molecularly-Detailed and Data-Driven Models

Introduction: Chimeric antigen receptor (CAR) T cells are T cells that have been engineered to express a receptor that binds to a tumor-associated antigen so that the T cells can detect cancer cells better than the normal immune system. While this treatment is successful in many patients, there are also many patients where this treatment is unsuccessful. It has been shown that protein levels can vary between cells of the same type, and that this can lead to phenotypically different cells within the same genotype. Since reaction kinetics depend on many proteins, it is difficult to analyze ODE-based mechanistic models to the fullest extent due to their high dimensionality. Data-driven models, while lacking the biological detail of a mechanistic model, are able to provide a more generalized relationship between model inputs and outputs. Here, we use a data-driven method, in the form of partial least-squares (PLS), to generalize the relationship of protein expression and cell response in CAR T cells. We used the mechanistic model to generate the large amount of data needed for this approach. Thus, the data-driven model is being used as an analysis tool for the mechanistic model. The results of this study should provide better understanding for the improvement of CAR therapy.

Materials and Methods: Using a mechanistic model of CAR T cell signaling, a heterogeneous population of 10,000 cells was simulated. Cell response was determined using the relative phosphorylated ERK concentration. This response separated the population into two categories: responders and non-responders. To better understand the relationship between protein expression and cell response, a PLS model was developed using the initial protein concentrations as inputs and the response as the output. The data generated from the mechanistic model simulations was used to train and validate the accuracy of the PLS model. Examining the components and the weights of the PLS model allowed us to determine the general influence of each individual protein on the cell's ability to respond. Since lower weighted inputs have less of an impact on the PLS outputs, proteins with lower weights were removed from the dataset and the PLS model was retrained to further determine which proteins are highly influential to the cell response.

Results and Discussion: From the population simulations, it was found that only about half of the cells in the population could be characterized as responders. A PLS model was developed that could classify this response with high accuracy. Out of the 19 proteins present in the network, only eight (LAT, Gads, SOS, Ras, RAF, MEK, RasGAP, and RasGRP) were found to be highly influential based upon the weights of their initial concentrations. The influence of these proteins on the system was found to match literature descriptions of their biological roles. Interestingly, it was found that the upstream proteins that interact with the CAR had very little influence on whether or not the cell would respond. All of the influential proteins identified are related to the activation of the MAPK pathway and its signal transduction. This approach of using a data-driven model as a way to analyze a mechanistic model allowed for a generalization of model parameters and investigation of their effects on the system as a whole. Additionally, we have used the data-driven model as a means of easing the computational burden of simulating the full mechanistic model. This enables a multi-scale framework to study immunotherapy, drug resistance, and the effects of tumor cell mutations.

Copyright © American Institute of Chemical Engineers. All rights reserved.

579608 Data Enabled Discovery in Soft Materials Discovery

The advent of new tools for ingestion and analysis of large volumes of data has created new and exciting opportunities for materials discovery and characterization. In the particular case of soft materials, including polymers, rapid advances are taking place on multiple fronts. In this presentation, I will discuss three examples, each of which serves to illustrate a different way in which data science can be used to extract detailed information about molecular processes or interactions. The first is concerned with the role of branching on polymer structure and dynamics, the second is concerned with development of coarse-grained models of conducting polymers, and the third focuses on the use of advanced sampling methods based on the use of concepts from machine learning.

Copyright © American Institute of Chemical Engineers. All rights reserved.

573067 Combining Molecular Dynamics Simulations and Deep Learning to Predict Interfacial Properties

Molecular dynamics (MD) simulations are a powerful tool for analyzing the properties and behavior of materials in solvent environments by explicitly representing atomistic degrees of freedom and accounting for intermolecular interactions. However, MD simulations can require lengthy runtimes that limit the high-throughput screening of system properties. Moreover, the analysis of MD simulation output typically requires the implementation of analytical techniques that rely on human insight to relate the properties of a statistical ensemble to relevant macroscopic properties. Alternatively, machine learning has emerged as a valuable tool to interpret molecular simulation data. Here, we present computational techniques that use machine learning to analyze molecular simulations of complex solvent environments. In particular, we focus on quantifying the hydrophobicity of functionalized interfaces by utilizing deep learning to analyze MD-derived interfacial water configurations. By selecting an appropriate representation of the MD output data, we demonstrate that a convolutional neural network (CNN) can infer spatial correlations between solvent molecules and map these features to thermodynamic observables. Once trained, we find that the CNN can predict dewetting free energies with minimal simulation input, facilitating the rapid screening of interfacial hydrophobicity for a range of surface compositions. We further use feature visualization methods to provide a physical interpretation of features identified by the neural network and compare these features to typical observables used to describe interfacial hydrophobicity (*i.e.*, water structure). This work demonstrates how deep learning methods can complement conventional MD simulations to facilitate the screening of interfacial properties.

Copyright © American Institute of Chemical Engineers. All rights reserved.

579611 Artificial Intelligence in Materials Science: The Good, the Bad, and the Ugly

The revitalization of artificial intelligence (AI) over the last decade due to the dramatic progress in data science has caused resurgence in the development of machine learning methods for materials science. The Holy Grail in materials science is to be able to design and/or discover new materials with desired macroscopic properties rationally, systematically, and quickly, rather than by the slow and expensive trial-and-error Edisonian approach that has dominated for decades. To accomplish this, one has to solve two different but related problems: (i) the forward problem, where one predicts the material properties given the structure or formulation, and (ii) the inverse problem, where one determines the appropriate material structure or formulation given the desired properties. AI methods have the potential to address this challenge in important ways.

However, applying AI to materials science is not new – it has a 35-year history and a rich literature. The forward-inverse conceptual framework, for instance, and its solution using hybrid neural networks and directed evolution, was demonstrated in 1992. What is exciting now is the ability to do all this more easily for more complicated materials due to the availability of powerful and user-friendly hardware and software environments, and, of course, plenty of data.

I classify the materials design problems into three categories – "easy", "hard", and "harder" classes. They correspond to the "good", the "bad", and the "ugly" problems, or the other way around, depending on your persuasion. The relatively "easy" ones are those where there is plenty of data that can be analyzed using off-the-shelf machine learning tools – for example, many structure-to-property prediction problems fall into this class. These can be, and are being, addressed right now. The "hard" problems are those which require combining first-principles knowledge of the underlying physics and/or chemistry with data-driven methods. Although researchers demonstrated how to do this a couple of decades ago, much work still remains to be accomplished to do this systematically and quickly for wider impact. This might take another decade or so.

Harder still is the last class, where one needs to develop domain-specific representations, languages, compilers, ontologies, molecular structure search engines, etc. – i.e., domain-specific "Watson-like" materials discovery engines. These really interesting and intellectually challenging problems would require going beyond purely data-centric machine learning, despite all the current excitement, and leveraging other knowledge representation and reasoning methods from the earlier phases of AI. They would require a proper integration of symbolic reasoning with data- driven processing. This might take a couple of decades to accomplish for extensive and routine usage. I will discuss these challenges and opportunities using materials design examples.

Copyright © American Institute of Chemical Engineers. All rights reserved.

570358 Multi-Objective, Machine-Learning Assisted First-Principles Design of Transition Metal Complexes for Redox Couples

Truly automatic, in silico design of materials with targeted properties is a central goal in the fields of chemistry, chemical engineering and materials science. This challenge is exacerbated by the enormous number of possible chemistries that could be considered, and by the fact that real materials must fulfil multiple criteria to be useful, e.g. we might seek redox couples that are active, stable and soluble. Unfortunately, many promising materials lie in relatively poorly explored regions of chemical space, for example the space of open-shell transition metal (TM) complexes, which are promising candidates for homogeneous catalysis and molecular electronics. While first-principles simulation with density functional theory (DFT) provides a probe to explore large, heterogenous design spaces, long simulation times and complicated electronic structure severely limits the number of candidates that can be considered. Machine learning (ML) methods can potentially address this problem by using the results of already-complete simulations to generate low-cost surrogate models for target properties, greatly increasing the number of candidates that can be screened. We have previously demonstrated that neural networks based on graph descriptors can be used to predict various properties of TM complexes to near baseline DFT uncertainty with a relatively small number (100s–1000s) of DFT simulations. However, when applying these models to tackle large design spaces, we observed highly variable generalization performance, reflecting the biases and limited variety of chemical motifs in the training data. Therefore, we have introduced geometric metrics to equip our surrogate models with reliable estimates of uncertainty, and we have been able to use these to avoid regions of low data coverage when designing spin crossover complexes and complexes with targeted orbital properties. Here, we use our uncertainty-equipped surrogate models to search a large, million-compound design space for TM complexes that have both target redox energetics and solubility properties for application in redox flow batteries. We use a multi-dimensional formulation of expected improvement that balances exploration (to improve model coverage) and exploitation (to find complexes with targeted properties) to screen this space for lead complexes that we then evaluate with DFT. By repeating this cycle, we can achieve first-principles coverage of the design space only where it is needed, iteratively refining a pareto set of candidate complexes that sample a range of redox and solvation properties.

Copyright © American Institute of Chemical Engineers. All rights reserved.

573946 Machine Learning the Fundamental Tradeoffs between Conductivity and Voltage Stability in Solid State Electrolytes

All solid state batteries provide many safety advantages over traditional lithium-ion batteries by replacing the combustible organic liquid electrolyte with a ceramic solid-state electrolyte. However, reported superionic conductors with conductivities approaching that of liquid electrolytes are unstable in contact with a lithium anode leading to increased internal cell resistance, and poor cyclability. Conversely, compounds stable at the anode or cathode interfaces do not exhibit useful bulk ionic conductivities. Although *ab initio* methods exist to study each ionic conductivity and voltage stability range, there is no established theory to connect these two properties. Here, we leverage machine learning (ML) to investigate the role of crystal structure in the tradeoff between voltage stability and ionic conductivity. To this end, we trained a partial least squares (PLS) machine learning algorithm using the valence electronic density as a descriptor of 60 known solid-state electrolytes along with their corresponding ionic conductivity, anodic voltage limit, and cathodic voltage limit. The trained model has an 80% prediction accuracy and suggests that within the search space of crystal structures, the voltage stability and ionic conductivity are inherently inversely correlated (Figure 1a). A multi-objective optimization also suggests that materials with positively correlated ionic conductivity and voltage stability may be highly anisotropic. Our PLS machine learning model, compared to the more conventional neural network, has the benefit of being able to predict and explore the relationship between multiple properties and retains a high level of interpretability versus other 'black box' models. The PLS model successfully identifies BCC anion substructure (Figure 1b) and channels as effective descriptors, which is in good agreement with prior work.[1] Using this model, we screened through a database of ca. 14,000 materials and identified five new promising solid state electrolyte candidates to have conductivities greater than 16 mS/cm. The model predictions were subsequently verified with *ab initio* molecular dynamics simulations. The proposed ML model and electron density descriptor may be used in future studies to elucidate structure-property relationships for other applications with high accuracy and without sacrificing interpretability.

Figure 1: (a) Latent variable space determined by the PLS model used to predict properties and (b) the BCC anion substructure identified by the PLS model.

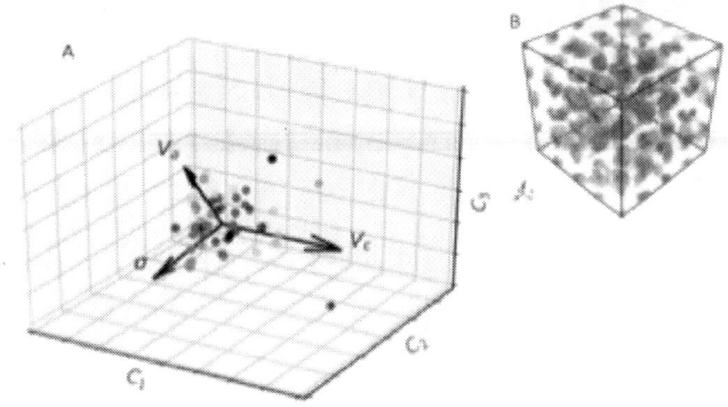

References:

1. Wang, Y. *et al.* Design principles for solid-state lithium superionic conductors. *Nat. Mater.* **14,** 1026–1031 (2015).

Copyright © American Institute of Chemical Engineers. All rights reserved.

576672 Machine Learning for the Prediction of Electronic Couplings and Charge Transport Calculations in P3HT

Predicting which organic chemical blends will make desirable solar cell materials requires a suite of computational chemistry tools including molecular dynamics simulations, quantum chemical calculations, and kinetic Monte Carlo simulations. In this work, we focus on ameliorating a quantum chemical calculation bottleneck in the pipeline between chemical ingredients and the charge mobilities that results from the self-assembled structure in these soft materials. Here we focus on the benchmark polymer poly(3-hexylthiophene) (P3HT) and calculating the dependence of charge mobility on P3HT nanostructure with kinetic Monte Carlo simulations that take charge hopping rates between chromophores as inputs. We compare five machine learning techniques for correlating structural features of P3HT monomers with the electronic couplings calculated with Zerner's method of intermediate neglect of differential overlap for spectroscopy (ZINDO/S). We find that random forests are an easy-to-implement, accurate predictor of couplings that finds the separation and orientation of bonded and non-bonded P3HT monomers chiefly determine their coupling. We discuss tests informing the minimal training set needed to generate random forests with 0.02 mean absolute error that is 400 times faster than performing new ZINDO/S calculations. We find that the charge mobilities predicted with the random-forest-generated couplings to be in agreement with the ZINDO/S mobilities. We conclude with an overview of the pipeline in which these random forests are used and how they are enabling broader screening of organic photovoltaic candidate chemistries.

Copyright © American Institute of Chemical Engineers. All rights reserved.

564718 Endotoxin Sensors Using Liquid Crystals and Machine Learning

Endotoxins are lipopolysaccharides that are found in the outer membranes of gram-negative bacteria and underlie many health hazards such as endotoxemia (a septic shock caused severe immune response) [1]. A number of methods have been proposed to detect the presence of endotoxins with remarkable sensitivity [2,3,4]. However, the most commonly used methods rely on complex reagents derived from the blood of crabs, do not provide a quantitative estimation of concentration, and/or are incapable of classifying endotoxin type.

Liquid crystal (LC) droplets dispersed in water are sensitive to the presence of endotoxins and undergo internal ordering transitions when exposed to different endotoxin species and concentrations [5]. Specifically, endotoxin accumulates at topological defects within the LC droplets, triggering a change from droplet configurations with bipolar symmetry to droplet configurations with radial symmetry. Different internal configurations of LC droplets have distinct optical properties [6]. As such, flow cytometry of LC droplets can be used to generate rich forward scattering/side scattering (FSC/SSC) data. A variety of counting methods used in the analysis of FSC/SSC data can predict endotoxin concentration [7], however, the optimal method for characterizing the scatter plots is not known and some methods are subject to variability.

In this work, we demonstrate that machine learning techniques can be used to accurately predict endotoxin concentrations and species from FSCC/SSC data. Our approach uses a convolutional neural network to extract pattern information (features) from FSC/SSC density data. Our framework reveals that significant hidden information is available in FSC/SSC data. Our analysis also provides insights into the physical effects driving changes in the LC internal configurations.

References:

[1] R. A. Proctor, *Handbook of Endotoxin*. Elsevier, 1984.

[2] T. Muta, T. Oda, and S. Iwanaga, "Horseshoe crab coagulation factor B. A unique serine protease zymogen activated by cleavage of an Ile-Ile bond.," *Journal of Biological Chemistry*, vol. 268, no. 28, pp. 21384–21388, 1993.

[3] K. G. Ong, J. M. Leland, K. Zeng, G. Barrett, M. Zourob, and C. A. Grimes, "A rapid highly-sensitive endotoxin detection system," *Biosensors and Bioelectronics*, vol. 21, no. 12, pp. 2270–2274, 2006.

[4] T. Y. Yeo *et al.*, "Electrochemical endotoxin sensors based on TLR4/MD-2 complexes immobilized on gold electrodes," *Biosensors and Bioelectronics*, vol. 28, no. 1, pp. 139–145, 2011.

[5] N. Bao, Y. Zhan, and C. Lu, "Microfluidic Electroporative Flow Cytometry for Studying Single-Cell Biomechanics," *Analytical Chemistry*, vol. 80, no. 20, pp. 7714–7719, 2008.

[6] I.-H. Lin, D. S. Miller, P. J. Bertics, C. J. Murphy, J. J. de Pablo, and N. L. Abbott, "Endotoxin-Induced Structural Transformations in Liquid Crystalline Droplets," *Science*, vol. 332, no. 6035, p. 1297, Jun. 2011.

[7] D. S. Miller, X. Wang, J. Buchen, O. D. Lavrentovich, and N. L. Abbott, "Analysis of the Internal Configurations of Droplets of Liquid Crystal Using Flow Cytometry," *Anal. Chem.*, vol. 85, no. 21, pp. 10296–10303, Nov. 2013.

Copyright © American Institute of Chemical Engineers. All rights reserved.

559792 Microscopic Modeling and Optimal Operation of Thermal Atomic Layer Deposition

Atomic layer deposition (ALD) process has been widely adopted by industry to produce thin-film materials that meet the requirements of major design breakthroughs in semiconductor industry [1]. Despite its popularity, the ALD process remains costly and it is desirable to develop an inexpensive high-fidelity model to characterize the process [2]. Recently, Shirazi and Elliot developed a Density Functional Theory (DFT)-based simulation of the HfO_2 deposition on laboratory scale [3]. Motivated by this work, a similar approach is expanded to develop a comprehensive framework for industrial-scale first-principles-based microscopic modeling, data-driven input/output modeling and optimal operation of thermal ALD of SiO_2 thin-films using bis(tertiary-butylamino)silane (BTBAS) and ozone as precursors [4].

In this work, we first perform first-principles-based DFT calculations of the key thermodynamic and kinetic parameters, which are then used in the microscopic modeling of the ALD process. Subsequently, a detailed microscopic model is constructed, accounting for the microscopic lattice structure and atomic interactions, as well as multiple microscopic film growth processes including physisorption, abstraction and competing chemical reaction pathways. Kinetic Monte-Carlo (kMC) algorithms are utilized to obtain computationally efficient microscopic model solutions while preserving model fidelity. The obtained kMC simulation results are used to train Artificial Neural Network (ANN)-based data-driven models that capture the relationship between operating process parameters and time to ALD cycle completion. Specifically, a dense two-hidden-layer feed-forward ANN is constructed to find a feasible range of ALD operating conditions accounting for industrial considerations, and a Bayesian Regularized ANN is constructed to implement the cycle-to-cycle optimization of ALD cycle time. Extensive simulation results demonstrate the effectiveness of the proposed approaches. The kMC models successfully achieves a growth per cycle (GPC) of 1.8 Å per cycle, which is in the range of reported experimental values. The ANN models accurately predict deposition time to steady-state from the given operating condition input, and the cycle time optimization using BRANN model reduces the conventional BTBAS cycle time by 60%.

Copyright © American Institute of Chemical Engineers. All rights reserved.

563078 Thermodynamic Properties Via a Combined Expanded Wang-Landau - Machine Learning Approach

Using machine learning (ML), we predict the partition functions and, thus, all thermodynamic properties of atomic and molecular fluids over a wide range of temperatures and pressures. Our approach is based on training neural networks using, as a reference, the results of a few flat-histogram simulations. The neural network weights so obtained are then used to predict fluid properties that are shown to be in excellent agreement with the experiment and with simulation results previously obtained on argon, carbon dioxide, and water. In particular, the ML predictions for the Gibbs free energy, Helmholtz free energy, and entropy are shown to be highly accurate over a wide range of conditions and states for bulk phases as well as for the conditions of phase coexistence. Our ML approach thus provides access instantly to G, A, and S, thereby eliminating the need to carry out any additional simulations to explore the dependence of the fluid properties on the conditions of temperature and pressure. This is of particular interest, for e.g., the screening of new materials, as well as in the parameterization of force fields, for which this ML approach provides a rapid way to assess the impact of new sets of parameters on the system properties. We then develop a combined Expanded Wang-Landau-Machine Learning (EWL-ML) approach to predict efficiently and accurately the thermodynamics of mixtures. The resulting speed-up is expected to considerably increase the range and complexity of systems that can be studied with even-sampling methods.

Copyright © American Institute of Chemical Engineers. All rights reserved.

569671 Understanding Mixing Behavior of Bimetallic Nanoparticles through Genetic Algorithm Modeling

Bimetallic nanoparticles (NPs) have received tremendous interest due to their controllable properties that make them promising candidates for a wide range of applications, such as catalysis and electronics. Central to these applications is to predict the thermodynamic preference of metals mixing at the nanoscale as a function of NP size, shape and metal composition. However, a complexity in bimetallic NPs is the explosion in the size of configurational space. In fact, it becomes impossible to study the entire search space of bimetallic NPs as their size increases, especially if one relies on first-principles calculations. To combat this problem, we develop a genetic algorithm (GA) capable of targeted screening towards the discovery of stable bimetallic NPs of any given morphology. Our GA leverages a previously reported "Bond-Centric" cohesive energy model[1] as a fast and accurate stability metric with practically minimal computational cost. With this approach we achieve high-throughput screening and can effectively search the large configurational space of bimetallic NPs. Using results from our GA, we connect the morphology, composition, and chemical ordering to the mixing behavior of bimetallic NPs – focusing on combinations of Cu, Ag, and Au. Importantly, we incorporate entropic effects into our framework to determine how mixing behavior changes with temperature. Taken together, our results unravel the impact of NP morphology (size, shape), metal composition, and temperature to the thermodynamic mixing behavior of bimetallic NPs and elucidates their detailed structural architectures (i.e. positioning of different metals on different NP sites).

1. Yan, Z.; Taylor, M. G.; Mascareno, A.; Mpourmpakis, G., Size-, Shape-, and Composition-Dependent Model for Metal Nanoparticle Stability Prediction. *Nano Lett.* **2018,** *18* (4), 2696-2704.

Copyright © American Institute of Chemical Engineers. All rights reserved.

573708 Estimation of Probability Density Functions of Model Parameters for a Heterogeneous Population with Neural Network: Application to TNFα Signaling Pathway

Modeling a heterogeneous population, whose dynamics are governed by same differential equations but with different parameter' values, has received much attention in various disciplines such as quantum physics and cellular biology [1-3]. Previous studies mostly focused on the problem of the optimal control of this heterogeneous population while methods on how to estimate such systems have begun to emerge [3-6].

Motivated by this, this study proposes a methodology to develop a heterogeneous dynamic model for a heterogeneous population. Specifically, we propose a numerical method to estimate the probability density functions (PDFs) of model parameters from experimental measurements. Here, we particularly consider the case where the output PDFs are available in the form of population snapshot data (PSD): PSD provide measurements from all members in the population at each time instance [6]. First, the PDFs of the model parameters are assumed to be normal so that the overall estimation problem become finite-dimensional. Second, Sobol' sensitivity analysis method is performed to determine a set of model parameters whose means and variances are practically identifiable. Third, a neural network model is developed to find an empirical function mapping from the PDFs of the identifiable parameters to the output PDFs. Lastly, the PDFs of the identifiable parameters are estimated by solving an optimization problem to minimize the difference between the measured and predicted output PDFs, which are computed based on PSD and the developed neural network model, respectively.

Our proposed methodology was implemented to infer the PDFs of the parameters in a tumor necrosis factor-α (TNFα) signaling pathway. Artificial PSD of the TNFα signaling dynamics were generated by simulating the model accordingly. Next, the proposed method was implemented to determine identifiable parameters and infer their PDFs from the PSD. After inferring the PDFs of these identifiable parameters, the corresponding output PDFs were calculated and compared with the PSD to validate the inference results. The reasonable agreement between the predicted and measured output PDFs demonstrated the validity of the proposed methods.

References

[1] Li, J.-S.; Khaneja, N. Control of inhomogeneous quantum ensembles. *Physical Review A* **2006**, *73*, 030302.

[2] Waldherr, S. Estimation methods for heterogeneous cell population models in systems biology. *Journal of the Royal Society Interface* **2018**, *15*, 20180530.

[3] Zeng, S.; Waldherr, S.; Allgöwer, F. Ensemble Observability of Linear Systems. *IEEE Transactions on Automatic Control* **2016**, 61, 1452-1465.

[4] Li, J.-S. Ensemble control of finite-dimensional time-varying linear systems. *IEEE Transactions on Automatic Control* **2011**, *56*, 345-357.

[5] Helmke, U.; Schönlein, M. Uniform ensemble controllability for one-parameter families of time-invariant linear systems. *System & Control Letters* **2014**, *71*, 69-77.

[6] Hasenauer, J.; Waldherr, S.; Doszczak, M.;Scheurich, P.; Radde, N.; Allgöwer, F. Analysis of heterogeneous cell populations: A density-based modeling and identification framework. *Journal of Process Control* **2011**, *21*, 1417-1425.

Copyright © American Institute of Chemical Engineers. All rights reserved.

567628 Development of Multimodal Cancer Diagnosis Method Based on Medical Image Analysis and Raman Spectroscopy

Digital pathology (DP) [1] is playing an ever-increasing role in clinical decision making. Technological advancements have made efficient data recording, processing, and analyzing possible. This enables the transformation of cancer diagnosis such that diagnosis can now be routinely assisted by algorithms in, for instance, computational image analysis [2], not least a wide variety of artificial neural networks (ANNs) developed for computer vision. Meanwhile, effective (and perhaps more intuitive) cancer identification and classification can also be realized based on appropriately selected morphological features of the biopsy samples, e.g. object contour, color, and texture [1]. It is therefore desirable to combine the advantages of both approaches so that a reliable and easy-to-interpret diagnosis can be achieved. Moreover, Raman spectroscopic analysis of specimens is gaining popularity in the biomedical research community [3], as molecular composition and microenvironment information of the cells unavailable in medical images can be highly sensitively captured in the Raman spectroscopy. Thus, Raman spectral profiles and their spatial distribution can be used to assist in characterizing cell/tissue samples. In this work, we construct a multimodal cancer diagnosis model based on the information provided by both the medical image analysis and Raman spectroscopy. Furthermore, we demonstrate the effectiveness and efficiency of the approach by running it on a wide variety of diagnostic samples.

[1] R. Bhargava and A. Madabhushi, *Annu. Rev. Biomed. Eng.* **18**, 387–412 (2016).

[2] D. Shen, G. Wu, and H.-I. Suk, *Annu. Rev. Biomed. Eng.* **19**, 221–248 (2017).

[3] H. J. Butler *et al.*, *Nat Protoc.* **11**, 664-687 (2016).

Copyright © American Institute of Chemical Engineers. All rights reserved.

572846 Invited: Sensor Applications in Bioengineering: Detection of Small Molecules within Complex and Three-Dimensional Biological Milieu

Bioengineering applications in regenerative medicine and single-cell analysis present challenging microenvironments for diagnostic and real-time detection. We have developed and applied sensors to detect small molecules in neuronal, cardiovascular, blood and biofilm systems. In particular, we have aimed to establish easy-to-use sensor platforms that do not interfere with biological processes. These sensors are reagentless and can be deployed in the presence of complex macromolecular solutions that readily foul, degrade or block many traditional sensing approaches. Herein, we describe our efforts towards detection of neuroactive molecules in the brain, hypoxia gradients in three-dimensional cellular scaffolds, and subcellular oxygen concentrations during electrophysiological and chemo-mechanical single cell experiments.

Copyright © American Institute of Chemical Engineers. All rights reserved.

575136 Tissue-like Neurochemical Sensors Operating in Brain and Gut

Normal neurotransmission relies on a precise and delicate balance of chemical signaling mediated by the release of variety neurotransmitters. Among them, the monoamines including dopamine and serotonin are uniquely involved in the regulation of cognitive processes such as emotion, arousal and memory. Therefore, it is of great interest in both basic research and clinical studies to monitor the fluctuation of monoamine in both brain and peripheral nervous systems. As a very soft and complex biomaterial, the human brain is under constant motion and deformation because of different physiological dynamics, while GI tract is a series of soft organs with inherent motility, joined in a long and twisting tube from the mouth to the anus. As a result, rigid neurochemical sensors can cause inflammation to the brain tissue, and neurochemical sensing in awake animals have historically been impossible to traditional rigid and fragile devices.

Softness and stretchability in new-generation neuroengineering tools are essential to minimize the perturbation of motions inherent in biological systems and maximize the physiological fidelity. Here we develop a new neuroengineering platform, based on soft materials, that could allow access to the brain and entire enteric nervous system for neurochemical sensing. The new neurochemical sensor is based on graphene-elastomer composite, which offers chronic, sensitive, and selective sensing of dopamine and serotonin in the brain. Ex vivo studies shows that the soft sensor could accommodate the peristalsis movement of the intestine and continuously monitor the serotonin dynamics via thermal treatment. In vivo experiments also show that the sensor could be implanted in the colon of mice to examine the drug-stimulated serotonin response. The new neuroengineering platform described here can provide unprecedented insights on the microbiota–gut–brain communication in the context of mental disorders and therapeutic interventions, and can lead to new therapies for psychiatric disorders through GI tract without invasive surgery to the brain.

Copyright © American Institute of Chemical Engineers. All rights reserved.

570825 Cell-Based Electrochemical Sensors for Environmental Pollutant Detection

Endocrine disrupting compounds are ubiquitous; they are found in many products intended for common use, including pesticides, plastics, and even some medications. Yet, these chemicals are implicated in health problems ranging from diabetes to cancer. We have developed a cell-based electrochemical strategy to rapidly and sensitively quantify these compounds. Our strategy measures the biological activity of chemically dissimilar xenoestrogens by harnessing the specificity of a human estrogen receptor. This receptor is expressed on the surface of *E. coli* as a component of an electrochemical sandwich assay. Our sensors require no specialized skills to implement and have enabled very sensitive detection (sub-ppb levels) of environmental pollutants. We have successfully detected compounds in from complex solutions, such as baby formula, and have measured estrogenic activity from unknown compounds leeched from a BPA-free baby bottle.

Copyright © American Institute of Chemical Engineers. All rights reserved.

577305 Biomimetic Composites As Microfabricated, Flexible, Degradable Electrochemical Sensors

Biomimetic composites as microfabricated, flexible, degradable electrochemical sensors

- M. Xu, S. Pradhan, R.K. Pal, V.K. Yadavalli

Department of Chemical and Life Science Engineering, Virginia Commonwealth University, Richmond, VA

Biomimetic composites of naturally derived and synthetic polymers provide exciting opportunities to develop mechanically flexible, physiologically compliant, and degradable bioelectronics systems. In prior work, we have shown that naturally derived silk proteins and conducting polymers (specifically poly(3,4-ethylenedioxythiophene) :poly(styrene sulfonate) (PEDOT:PSS)) can form functional bioinks for sensing and other electrochemical applications. These inks combine the mechanical strength, biocompatibility and degradability of silk proteins, with the conductivity and chemical functionalization of the PEDOT:PSS. The fabrication and characterization of photopatternable, water-based conductive inks comprising PEDOT:PSS and synthesized photoreactive silk proteins was earlier shown [1] The presence of photoreactive groups permits a fully aqueous photolithographic strategy to form conductive microelectrodes on both rigid substrates as well as flexible silk films [2].

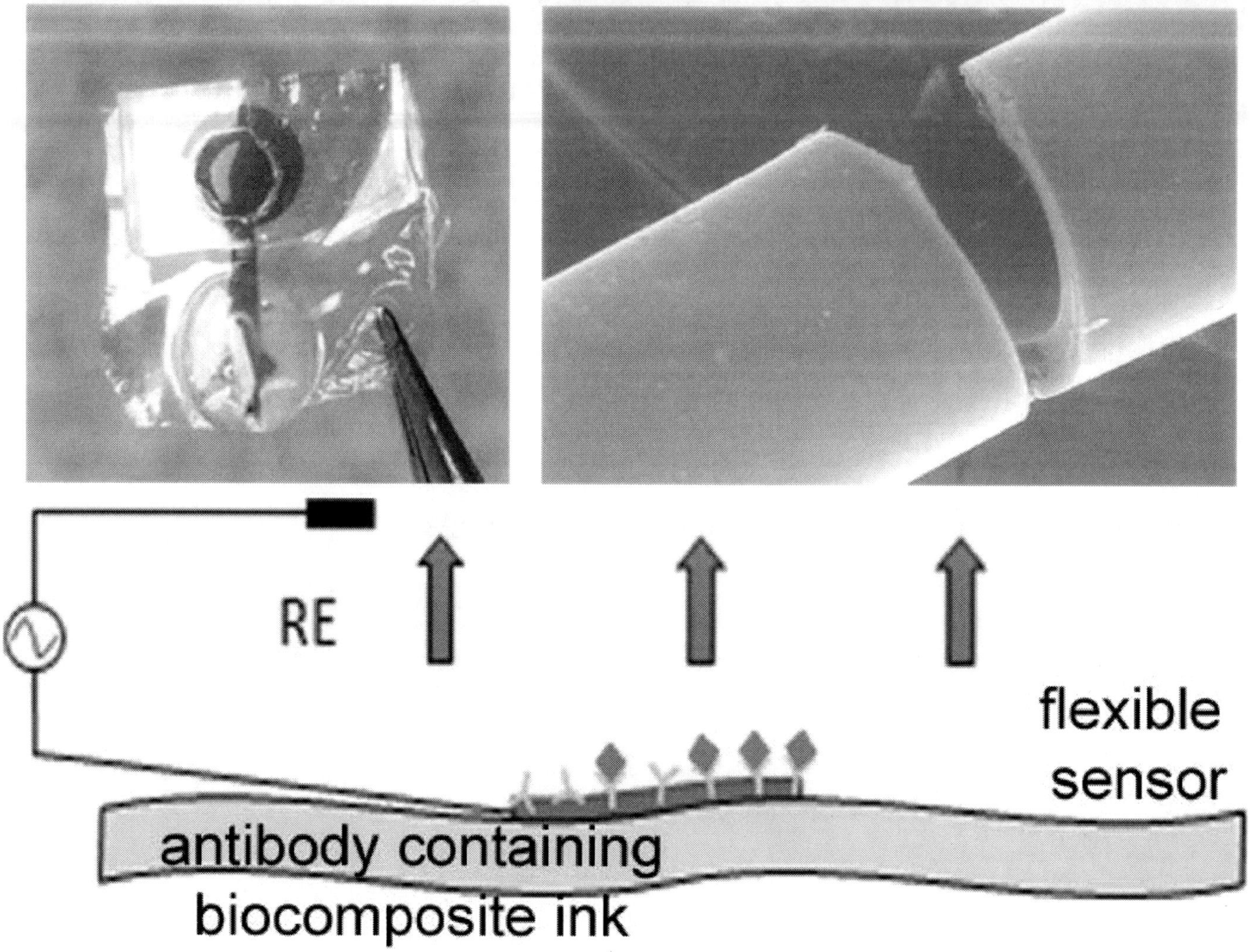

Here we will present the latest developments using this technology, showing how these devices can function for flexible bioelectronics applications in the form of microelectrodes for ultrasensitive detection, as well as components such as microsupercapacitors.[3, 4] We show how electroactive biomolecules such as neurotransmitters can be detected sensitively, while non-electroactive biomolecules such as glucose and glutamic acid, can be detected by encapsulating specific enzymes. Due to the room temperature, aqueous fabrication process, we can immobilize antibodies in these sensors resulting in flexible immunosensors for targets such as vascular endothelial growth factor (VEGF) for the first time [3]. The electroactivity of conductive ink can be improved by the addition of small amounts dopants. Using these doped composites, we further demonstrate flexible energy storage devices due to the capacitive nature of the biomaterial. The presence of silk proteins as the matrix of the composite makes it completely biodegradable and biocompatible, potentially resulting in transient or implantable devices. This platform can also be used for the specific control of cell culture in flexible sheets. The mechanical, biochemical and electrochemical characterization of the composite and its microfabrication are discussed. By virtue of a range of versatile properties, utility as bio-sensors, opto-electronic devices and flexible energy storage systems are envisioned.

References:

[1] "Photolithographic micropatterning of conducting polymers on flexible silk matrices" - RK Pal, AA Farghaly, MM Collinson, SC Kundu, VK Yadavalli, *Advanced Materials*, 28(7), 1406-1412, 2016

[2] "Conducting polymer-silk biocomposites for flexible and biodegradable electrochemical sensors" - RK Pal, AA Farghaly, MM Collinson, SC Kundu, VK Yadavalli, *Biosensors & Bioelectronics*, 81, 294-302, 2016

[3] "Flexible biosensors for the impedimetric detection of protein targets using silk-conductive polymer biocomposites" – M Xu, VK Yadavalli, *ACS Sensors*, in press, 2019

[4] "Fabrication of Flexible, Fully Organic, Degradable Energy Storage Devices Using Silk Proteins" - RK Pal, SC Kundu, VK Yadavalli, *ACS Applied Materials & Interfaces* 10 (11), 9620-9628, 2018

Copyright © American Institute of Chemical Engineers. All rights reserved.

569635 Nanopore Based Highly Sensitive and Accurate Detection of HIV

Nanopore Based High Performance Point-of-Care Biosensor for Acute HIV Detection

Xiaojun Wei, Chang Liu

Human immunodeficiency virus (HIV) is the causative agent of acquired immuno-deficiency syndrome (AIDS). There are over 1.8 million infections HIV occur annually worldwide, and 30%¨C50% are attributed to recently infected individuals that are unaware of their disease state. Rapid diagnosis of acute HIV (within 3 weeks of infection) may reduce new cases by 10%¨C25%. The fourth-generation rapid diagnostic test for HIV is currently employed worldwide as the frontline screening test. This Enzyme-linked Immunosorbent Assay (ELISA) based test is designed to detect HIV-1 p24 antigen in addition to HIV antibodies to achieve early detection. However, several recent field tests have indicated unacceptably low diagnostic sensitivity (0%) of acute HIV-1.

Herein, a nanopore biosensing platform with single-molecule level sensing and analysis capacity has been developed to address the unmet need for rapid and accurate diagnosis of acute HIV to reduce transmission. In our approach, p24 antigens isolated from plasma samples were cleaved into several specific peptide sequences, which can be recognized and quantified by their signature signals during translocation through the nanopore. A portable QuantiPep-HIV prototype suitable for point-of-care settings was developed and tested with serum samples from HIV-infected and uninfected individuals to demonstrate the feasibility and to evaluate the diagnostic performance. Our results showed that QuantiPep-HIV can improve the diagnostic sensitivity and specificity of acute HIV.

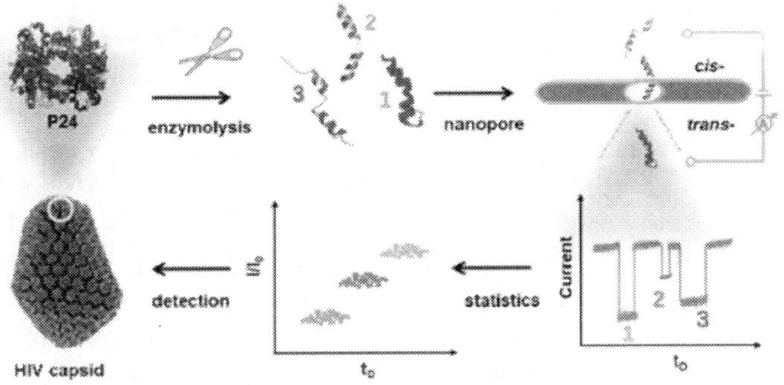

Figure. Schematic diagram for detecting peptide fragments from p24 antigens by nanopore

Copyright © American Institute of Chemical Engineers. All rights reserved.

582378 Wireless, Soft, Skin-integrated Microfluidic Systems for Electrochemical/Colorimetric Biosensing and Capture of Sweat

Human eccrine sweat is a rich composition of nearly 60 known biochemicals including metabolites, electrolytes, vitamins, amino acids, exogeneous drugs, and small proteins. Advances in sensing technologies, conformal electronics, and materials science lay the foundation for recent demonstrations of wearable sweat sensors. Unfortunately, majority of the devices rely either on active, battery-powered electronics for electrochemical detection or passive, colorimetric chemistries for visual readout. Complex construction, large size/weight and possibility of battery leakage represent disadvantages of the former; semi-quantitative operation and limited scope of measurable biomarkers are drawbacks of the latter. Additionally, present wearable sensors can detect only a handful of biomarkers present in the sweat. Quantification of the remaining chemicals still requires sweat collection using adsorbent pads followed by off-site chemical analysis. Such a route provides average concentration of sweat analytes and restricts the acquisition of their temporal concentrations. This prohibits the understanding of the relation between dynamic variations in sweat chemistry and human physiology. In this talk, I will describe an unusual collection of enabling materials, sensing technologies, advanced microfluidics, stimulus dependent biocompatible power systems, galvanic time-tracking schemes, and wireless electronics to address these issues. Specifically, I will describe my strategies that include soft microfluidics, fuel cells, trigger-activated biocompatible batteries, colorimetrics and near field communication technologies that result in new classes of wireless, skin-integrated devices for detailed analysis of human sweat.

Copyright © American Institute of Chemical Engineers. All rights reserved.

574301 The Electrochemical Impedance Response of a Continuous Glucose Monitor

The Gerischer impedance has been observed in systems with an electrochemical reaction coupled with a homogeneous chemical reaction. Under the assumptions of linear homogeneous reaction, equal diffusion coefficients and a semi-infinite condition, the Gerischer impedance can be solved analytically [1, 2]. However, with non-ideal electrochemical systems, where coupled nonlinear homogeneous reactions and various mass-transfer properties make exact analytical resolution impossible, a more detailed analysis is required.

In the present work, a mathematical model for the impedance response of glucose-oxidase based electrochemical biosensors has been developed [3]. The complicated coupling between thirteen homogeneous reactions and three heterogeneous reactions has been considered in the model. The homogeneous reactions included anomerization between α-D-glucose and β-D-glucose and four reversible enzymatic catalytic reactions transforming β-D-glucose and oxygen into gluconic acid and hydrogen peroxide, pH-dependent enzymatic activity and a biological buffer system. The electroactive hydrogen peroxide was considered to be reversibly oxidized or reduced at the electrode.

The mathematical model is solved numerically by using the finite-difference method and Newman's BAND algorithm [4]. The model demonstrates how the coupled non-linear homogeneous reactions effect the diffusion impedance, which has broadened the scope of the Gerischer impedance. The model can be used to explore the influence of various system parameters on limiting current, reaction profiles, and diffusion impedance. The system parameters, including interstitial glucose concentration, oxygen concentration, active enzyme concentration, diffusion coefficients, reaction rate constants and layer thickness, are related to various sensor working conditions such as body sugar level, inflammation, sensor degradation and sensor design.

References

1. H. Gerischer, "Wechselstrompolarisation Von Elektroden Mit Einem Potentialbes- timmenden Schritt Beim Gleichgewichtspotential," *Zeitschrift fur Physikalische Chemie*, **198** (1951) 286–313.
2. M. E. Orazem and B. Tribollet, *Electrochemical Impedance Spectroscopy* (John Wiley & Sons, Hoboken, NJ, 2017), 2nd edition, p. 279-293.
3. M. Gao, M. S. Hazelbaker, R. Kong, and M. E. Orazem, "Mathematical Model for the Electrochemical Impedance Response of a Continuous Glucose Monitor," *Electrochimica Acta*, **275**(2018),119-132
4. J. S. Newman and K. E. Thomas-Alyea, *Electrochemical Systems* (John Wiley & Sons, Hoboken, NJ, 2004), 3rd edition.

Acknowledgement

The support of Medtronic Diabetes (Northridge, CA) and Andrea Varsavsky, program monitor, is gratefully acknowledged.

Copyright © American Institute of Chemical Engineers. All rights reserved.

560000 Impact of Uncertainty Quantification and Density Functional Theory Error Propagation on Surface Catalyzed Kinetic Models

Parametric uncertainty currently limits the predictive ability of kinetic models and prevents them from reaching their full potential. This is particularly important for surface catalyzed microkinetic models which often contain thousands of reaction steps and intermediate surface species requiring many kinetic and thermodynamic parameters. Estimating these from first-principles density functional theory (DFT) is costly, motivating the need for DFT-based methods, such as group additivity and Brønsted–Evans–Polanyi (BEP) relationships, to quickly estimate thermochemistry and reaction barriers, respectively. Uncertainty in DFT and these methods is amplified by the Boltzmann factors in the reaction rate expressions increasing uncertainty in estimating the overall reaction rate. Furthermore, microkinetic model parameters are correlated; yet, there is lack of a framework to expose these correlations on a single catalyst with a single functional. This work develops a framework to estimate correlations in reaction networks and DFT uncertainty propagation. Uncertainty quantification (UQ) is then predicted for thermochemistry, reaction barriers, reaction path, and ultimately reaction rates. Propane combustion and ethane oxidative dehydrogenation reactions are illustrated. We will also demonstrate how to use the Python Multiscale Thermochemistry Toolbox, pMuTT [1], to calculate uncertainty in apparent activation energy.

References

[1] Python Multiscale Thermochemistry Toolbox (pMuTT), https://pypi.org/project/pMuTT/

Copyright © American Institute of Chemical Engineers. All rights reserved.

568912 Automated Machine Learning and Microfluidics-Based Study and Optimization of Zeigler-Natta Catalysts

Development of new polymerization processes and the requisite catalysts is a complex task, often taking years and the combined efforts of multiple people, especially when optimizing for various parameters such as optimal chain length, polydispersity index, and economic factors. This challenge has only increased in recent years as there has been a move toward homogeneous Ziegler-Natta catalysts, a class of molecules which provides favorable control and processing characteristics but are notoriously difficult to optimize and understand. Optimization of a Ziegler-Natta catalyst system relies on the careful selection of several key parameters including the solvent, type and concentration of activator, catalyst concentration, monomer addition, temperature and mixing. While all these factors play into generating the desired polymer, the relationship between these variables is often not clear due to the chemistry involved. Proposed mechanisms for this reaction include up to eleven different equilibrium and non-equilibrium reaction steps, the careful balance of which effects the properties of the final polymer. Traditionally this has been reconciled by performing combinatorial experiments to individually judge the contribution of each factor, but these experiments take time and create large amounts of chemical waste. In our research we aim to apply new laboratory techniques including high throughput microfluidic screening and machine learning to screen, understand and optimize these catalysis systems.

By miniaturizing the system and performing the experiment in a microfluidic system we are able to carefully control the flow properties and temperature gradients and tie a key characteristic of the reaction- the exotherm- to the activity of the catalyst. The critical step in the polymerization mechanism is the propagation of the polymer chain where the pi-bond in an alpha-olefin is converted to a sigma bond, releasing large amounts of energy. By monitoring this progression in a microchannel using IR thermography we are able to get a real time idea as to the concentration of monomer left in the system and thus quantify the activity of the catalyst under various conditions. This is supplemented with measurements by a Dynamic Light Scattering (DLS) instrument to quantify the chain length of the polymers, an important industrial characteristic. Overall this microfluidic platform allows us to capture relevant kinetic information in a highly controlled and repeatable fashion.

The second component which comes into play for the fast screening and optimization of these polymerization catalysts an automated reagent delivery system and control software. The reagent delivery system is constructed in such a way as to allow for a continuously variable composition of reactants- necessary for understanding broad reaction spaces. Information from the experiment is fed back to a machine learning algorithm trained on first-principles data which interprets the activity of the catalyst and updates the training. Based on the updated training the algorithm predicts a new theoretical maximum efficiency range for the catalyst and sends that particular composition to the pumping sub-system to perform the experiment. Overall this allows for the fast screening of catalysts and for the derivation of optimal operating conditions without the need for numerous manual experiments.

Overall while this methodology is currently being applied to the academic study of Zeigler-Natta catalysts, the overall goal is to enhance industrial adoption of AI and ML for the development of chemical processes. By being able to integrate highly multidimensional data and derive optimized conditions automatically we hope to enable a new path towards catalyst discovery.

Copyright © American Institute of Chemical Engineers. All rights reserved.

576554 Simultaneous Canonical Polyadic Decomposition As a Data Fusion Algorithm to Develop Pseudo-Chemistry from Spectral Data

Simultaneous Canonical Polyadic Decomposition as a Data Fusion Algorithm to Develop Pseudo-chemistry from Spectral Data

Keywords: Data Mining, Tensor Decomposition, Data Fusion, Bayesian Networks, Reaction pathway generation

In-line spectral analyzers are popularly used to obtain molecular-level information as they are fast, non-invasive, non-destructive, inexpensive and do not require sample preparation. The process data from spectral analyzers are high dimensional, multi-way, non-causal, non-full rank and have missing values. This offers a challenge in using such process data to develop causal models that hypothesize reaction pathways. This work uses the spectral datasets from Fourier Transform Infrared (FTIR) and Proton Nuclear Magnetic Resonance (^{1}HNMR) spectroscopy from the vis-breaking process of Cold Lake bitumen. The absorbance data over wavenumber/ chemical shifts are collected across different temperatures and residence time of reaction of the samples in the vis-breaker. The two spectral measurements collected across the three modes viz. temperature, residence time and absorbance/ chemical shifts represent multi-way tensorial blocks of process data. The objective of this work is to develop a data fusion framework that does not discount the tri-linear structure of the tensorial blocks of spectral measurements, accounts for the missing spectral measurements and the high-dimensionality of absorbance data across wavenumbers/ chemical shifts while aiming to incorporate complementary information about the sample from both. This is accomplished using the scheme of simultaneous canonical polyadic decomposition (CPD) where the tensorial blocks of spectral data are jointly factorized into independent factors in each mode while capturing intermodal interactions during the decomposition, resulting in a unique factorization scheme that is free from rotational and intensity ambiguities [1].

The jointly mined process data using the data fusion algorithm is then used to develop inferential models for monitoring the complex process of vis-breaking by developing pseudo-reaction networks that hypothesize chemical pathways [2]. It is to be noted that the difficulty in analytical characterization of the reacting mixture means that even the enumeration of all the major species taking part in reactions is a significant challenge for conventional methods. This is done using a probabilistic framework wherein the independent factors of a mode obtained from the simultaneous CPD are viewed as random variables with a multinomial distribution (the hyperparameters of which have a Dirichlet distribution). Bayesian networks, which are probabilistic graphical models that encode directed acyclic causal structures among the nodes of these random variables are used to develop inferential hypotheses about chemical pathways from the factors obtained from data fusion of spectral measurements. A directed path exists between nodes if it maximizes the log likelihood, which is a function of the mutual information and entropy, calculated using the probability distributions of the random variables designated as nodes (factors obtained from CPD). This amounts to using a score called the Bayesian information criterion (BIC), which is the log likelihood of the entire network (pairwise directed edges between nodes) penalized by the complexity of the network (number of edges between nodes). Heuristic greedy search score-based methods that make locally optimum choices while checking to see if directed edges between pairwise nodes maximize the penalized log likelihood are used to obtain the Bayesian networks i.e. the directed acyclic graph (DAG) encoding causal relationships among the factors obtained from simultaneous CPD. During data fusion using the simultaneous CPD, the independent factors in each mode are constrained to be non-negative so that the decomposition is physically meaningful by complying with the Beer Lambert law for spectral data, which states that the absorbance is directly proportional to the concentration of the components. Hence the independent factors in each mode from the data fusion algorithm can be physically interpreted as representing a class of chemical compounds (pseudo-component). The trilinear tensorial decomposition helps us obtain the concentration of these pseudo-components in the modes of temperature and residence time; while the third mode of wavenumbers/ chemical shifts contain the spectral signatures of the corresponding pseudo-components.

The number of independent factors in CPD is obtained using an important diagnostic called the core consistency diagnostic [3], whereby the number of factors in a CPD is obtained by fitting PARAFAC models with arbitrary factors and then casting them into Tucker models, which should produce an identity hypercube if the right number of components are used. This technique exploits the fact that tensor decompositions are higher order extensions of matrix singular value decompositions (SVD) and broadly fall into two categories: Parallel factor decomposition (PARAFAC), which represents a tensor as a sum of several rank 1 tensors, and Tucker decomposition, which is a higher order SVD. The main difference between CPD using PARAFAC and Tucker is that the number of factors is invariant across the modes in the former, making it a restricted version of the latter. When this diagnostic was used on the tensor blocks of FTIR and ^{1}HNMR data in this work, the number of factors obtained was four. The spectral signatures of the four pseudo-components obtained from the data fusion algorithm (Fig. 1) were then used to develop Bayesian networks (Fig. 2).

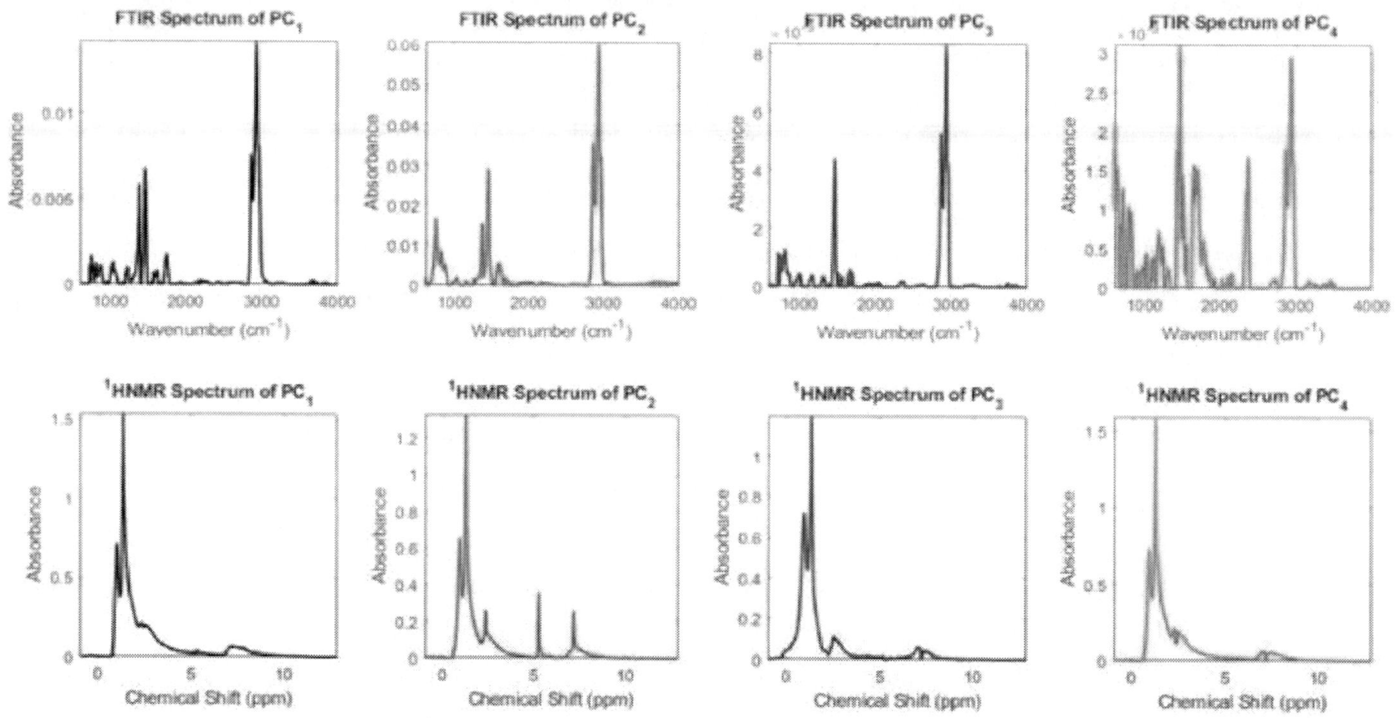

Figure 1. Spectral signatures of pseudo-components obtained from the data fusion algorithm

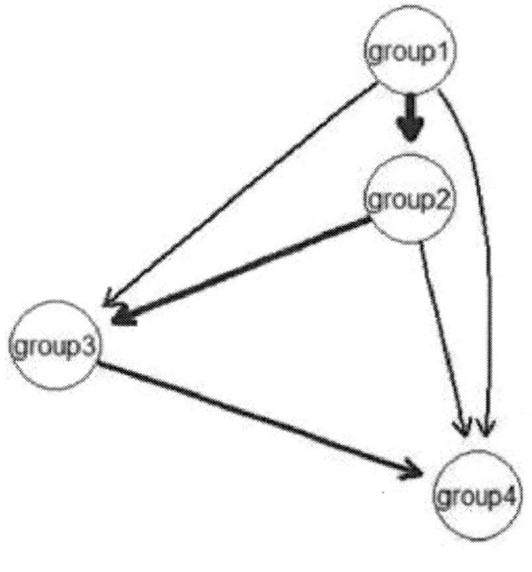

Figure 2. Bayesian networks developed from the spectral signatures of the pseudo-components (PC)

The analysis of the spectral signatures reveal the following: pseudo-component 1 (PC_1) mainly consists of carbonyl groups and cycloalkanes; pseudo-component 2 (PC_2) consists of polyaromatics, alkoxy groups, phenols, alkenes; pseudo-component 3 (PC_3) consists of aromatics, alkanes and condensed products; and pseudo-component 4 (PC_4) consists of phenols, acyls and condensed aromatics. It can therefore be hypothesized from the above Bayesian network structure that the underlying chemical reaction pathways during the vis-breaking of bitumen aim at obtaining more saturated end products through the free radical mechanism of hydrogen radical addition. However, the longer chain aliphatics crack to give more condensed polyaromatic products which are undesirable even as the end products of thermal cracking have a more aliphatic nature (alkanes and olefins).

The novelty of this work lies in the implementation of a constrained simultaneous CPD of multiple tensors using an optimization approach [4] of solving for the decision variables (factors) using the gradients of the objective function, which is formulated as the reconstruction error of the tensors from their multi-modal factors. This is an improvement over the Canonical Polyadic Alternating Least Squares (CPALS) approach, typically used for the unconstrained decomposition of just one tensor block. CPALS is not very accurate as it is not guaranteed to converge to a stationary point. Besides, the simultaneous CPD developed in this work is designed to handle missing data by imputing them by a weighting matrix in the objective function of the optimization framework. The causal DAG among the factors obtained from jointly decomposing both the tensors (FTIR and ^{1}HNMR measurements) is developed using Bayesian networks and is representative of the underlying chemical pathways among the pseudo-components. This work facilitates jointly mining spectral measurements in the framework of constrained data fusion to make the factors physically interpretable so that a first pass to building causal inferential models to generate reaction network hypotheses from process data (spectral measurements) could be implemented.

References

[1] T. G. Kolda and B. W. Bader, "Tensor Decompositions and Applications ," *SIAM Rev.*, vol. 51, no. 3, pp. 455–500, 2009.

[2] D. T. Tefera, L. M. Yañez Jaramillo, R. Ranjan, C. Li, A. De Klerk, and V. Prasad, "A bayesian learning approach to modeling pseudoreaction networks for complex reacting systems: Application to the mild visbreaking of bitumen," *Ind. Eng. Chem. Res.*, vol. 56, no. 8, pp. 1961–1970, 2017.

[3] R. Bro and H.A. Kiers, "A new efficient method for determining the number of components in PARAFAC models," *J.Chemometrics*, Vol.17,274-286, 2003.

[4] E. Acar, D. M. Dunlavy, and T. G. Kolda, "A scalable optimization approach for fitting canonical tensor decompositions ," vol. 25, no. 2, pp. 67–86, 2011.

-

-

-

Copyright © American Institute of Chemical Engineers. All rights reserved.

569295 Bridging the Synthesis-Surface Information Gap: Incorporating Spectroscopy Data into an Experimental Machine Learning Framework

One current challenge in machine learning for heterogeneous catalysis is selecting descriptors which correctly relate catalyst composition to the surface properties that directly influence activity. For machine learning to be predictive, it requires that all features input to the model are easily accessible or calculable, but collection of surface information requires measurement of synthesized catalysts. This often precludes using surface information as a predictor for catalyst activity, since the goal of a catalyst discovery framework is to predict activity without needing to synthesize new catalysts. To bridge this gap, a machine learning model is created to encode Raman spectroscopy data into spectroscopic descriptors. This model is trained to predict these descriptors from easily accessible catalyst properties, such as electronegativity or ionization energy, and experimental synthesis conditions. A similar methodology has been demonstrated for x-ray diffraction data but has yet to be applied to surface sensitive techniques such as Raman spectroscopy [1]. These intermediate surface descriptors are subsequently used alongside easily accessible catalyst properties as training data in a machine learning algorithm to predict catalyst activity for unknown catalyst formulations.

[1] A. Corma, J. M. Serra, P. Serna, M. Moliner, Integrating high-throughput characterization into combinatorial heterogeneous catalysis: Unsupervised construction of quantitative structure/property relationship models. *J. Catal.* **232**, 335–341 (2005).

Copyright © American Institute of Chemical Engineers. All rights reserved.

570431 Machine Learning and Transition Metal Chemistry: Data-Driven Comparisons of First and Second Row Complexes

The design of molecular transition metal (TM) complexes with targeted properties has great potential for homogeneous catalysis, molecular electronics and functional materials. However, the space of possible TM complexes is enormous and poorly understood, especially compared to organic chemistry where notions of chemical similarity are well established. High-throughput virtual screening of open-shell TM complexes by density functional theory can estimate the properties of novel complexes but is complicated by spin and oxidation state dependent behavior, long computation times and difficulty in reliably converging to the intended geometries and electronic states. To address these challenges, we have developed a unified strategy for the data-driven analysis and design of transition metal complexes, combining autonomous simulation, specifically engineered geometry-free representations, and fast property prediction with neural networks, uncertainty estimation and machine-learning informed workflow control. By exploiting a few thousand DFT calculations on first row TM complexes, we were able to train neural networks and use them to design new complexes with targeted spin state ordering and frontier orbital properties. By directly incorporating measures of model confidence, we are able to validate most leads at a DFT level, providing at least an order-of-magnitude increase in design efficiency compared to DFT alone. Additionally, we predict the likelihood of success of calculations to avoid performing unproductive simulations based on live analysis of the wavefunction, greatly increasing the success rate of screening and avoiding waste of computational resources. Now, we apply our automated framework to investigate properties of second-row TM complexes containing centers Mo-Rh, making a comparison to the electronic and geometric properties first-row equivalents. We assess transferability of models trained on first row data to this space and find that our outcome prediction model is transferable and can effectively reduce the number of unproductive simulations on these systems. Model transferability for property prediction is a strong function of the representation choice, and representations that directly encode periodic information can lead to improved learning both between rows and on out-of-sample errors for each row when re-training on the combined data. This demonstrates how data-driven methods can be used as a lens to examine chemical similarity and identify chemical trends in design spaces that are otherwise relatively poorly understood.

Copyright © American Institute of Chemical Engineers. All rights reserved.

570788 Active Machine Learning Model for the Discovery of Novel Material Systems and Its Application to Electrocatalysis

The design and discovery of new material systems with desired chemical and physical properties is imperative towards addressing global energy challenge. Traditional paradigms in material development are often slow, labor intensive and are not guaranteed to fully explore the expansive chemical space. Herein, we report on a robust and generalizable active learning framework for the discovery of novel and stable bulk crystal structures in the full space of crystal structural prototypes of fixed stoichiometry. Furthermore, we demonstrate the utility of this framework by the search in space of Ir-oxides, which has direct relevance to materials application as oxygen evolution reaction (OER) and oxygen reduction reaction (ORR) catalyst and support.

An active surrogate machine learning algorithm employing a Gaussian kernel was implemented to perform a rapid exploration of the Ir-oxide chemical space. The structural space was populated by all unique AB_2 and AB_3 structures from Materials Project and Open Quantum Materials Database (OQMD) databases using the prototype generator of Jain, et al available at https://www.catalysis-hub.org/prototypeSearch [1][2]). Since this approach can be easily generalized to rigorously enumerate a full structural space it also guarantees the completeness of our search. We show that the active learning framework can select the most valuable structures to perform DFT calculations on by implementing a simple acquisition criteria that biases towards low energy structures. The surrogate model resolves the most stable materials in the target space with a minimal number of DFT calculations (enough for the model's accuracy in the low formation energy region to be sufficiently low). Next, the subset of most stable IrO_2 and IrO_3 structures was computationally tested for their efficacy towards the electrochemical OER. The results for our bulk and surface calculations show the importance of highly oxided Ir-phases and provide atomistic explanation of the observed high stability and activity of the previously reported IrO_x/Sr_xIrO_3 system [3]. This work demonstrates the importance of data science approaches to accelerate catalyst discovery.

References

[1] Jain A., Bligaard T, Physical Review B **2018**, 98, 214112.

[2] Winther K., Hoffmann M., Mamun O., Boes J., Nørskov J., Bajdich M., Bligaard T., ChemRxiv **2018**,

[3] Seitz L., Dickens C., Nishio K., Hikita Y., Montoya J., Doyle A., Kirk C., Vojvodic A., Hwang, H., Nørkov J., Jaramillo T., **2016**, 353, 6303

Copyright © American Institute of Chemical Engineers. All rights reserved.

572243 A Machine Learning Framework for the Analysis and Prediction of Catalytic Activity from Experimental Data

We present a machine learning framework to explore the predictability limits of catalytic activity from experimental descriptor data that characterizes catalyst formulations and reaction conditions. Artificial neural networks are used to fuse descriptor data to predict activity and we use principal component analysis (PCA) [1] and sparse PCA [2] to project the experimental data into an information space and with this identify regions that exhibit low- and high-predictability. Our framework also incorporates a constrained-PCA optimization formulation that identifies new experimental points while filtering out regions in the experimental space due to constraints on technology, economics, and expert knowledge. This allows us to navigate the experimental space in a more targeted manner. Our framework is applied to a comprehensive water-gas shift reaction data set, which contains 2,228 experimental data points collected from the literature [3]. Neural network analysis reveals strong predictability of activity across reaction conditions (e.g., varying temperature) but also reveals important gaps in predictability across catalyst formulations (e.g., varying metal, support, and promoter). PCA analysis reveals that these gaps are due to the fact that most experiments reported in the literature lie within narrow regions in the information space. We demonstrate that our framework can systematically guide experiments and the selection of descriptors in order to improve predictability and identify new promising formulations.

[1] Jolliffe, Ian. *Principal component analysis*. Springer Berlin Heidelberg, 2011.

[2] Zou, Hui, Trevor Hastie, and Robert Tibshirani. "Sparse principal component analysis." *Journal of computational and graphical statistics* 15.2 (2006): 265-286.

[3] Odabaşı, Çağla, M. Erdem Günay, and Ramazan Yıldırım. "Knowledge extraction for water gas shift reaction over noble metal catalysts from publications in the literature between 2002 and 2012." *International Journal of Hydrogen Energy*39.11 (2014): 5733-5746.

Copyright © American Institute of Chemical Engineers. All rights reserved.

573779 Graph Convolutional Machine Learning Methods for the Predictions of Adsorption and Thermochemistry and Surface Stability

Prediction of surface phenomena such as surface stability or adsorption thermochemistry in catalysis have traditionally focused on simple coordination-based models or correlated these properties with electronic descriptors such as d-band properties. In recent work we showed that flexible coordination based fingerprints with sufficient data (thousands of calculations) were sufficient to achieve ~0.2 eV MAE accuracy for CO/H adsorption energies across a wide range of catalyst compositions. In this work we show that graph convolutional methods that learn important features and local activity from local atomic environments outperform our previous approach and are capable of achieving ~0.14 eV accuracy across thousands of surfaces with >30 unique elements while simultaneously predicting the contribution of each surface atom to the overall property. We also demonstrate this approach for a new dataset of intermetallic surface energies and show that these methods are capable of qualitatively predicting dominant nanoparticle facets. This works for sufficiently large datasets and demonstrates the transition from small-data (featurization and machine learning) to large data (deep learning) approaches for the prediction of surface phenomena in catalysis.

Copyright © American Institute of Chemical Engineers. All rights reserved.

571027 Data-Driven Design of Self-Assembling Pi-Conjugated Oligopeptides (Invited Talk)

Synthetic oligopeptides containing pi-conjugated cores present an attractive molecular building block for the self-assembly of biocompatible nanoaggregates with engineered optical and electronic properties. The peptidic and pi-conjugated blocks can be tailored to control the aggregation morphology and emergent optoelectronic properties such as electronic delocalization, electron/hole transport, and optical absorption spectrum. The design space of possible peptide sequences and pi-conjugated cores is so large as to make Edisonian trial-and-improvement of the structure and properties of the self-assembled aggregates intractable by either experimentation or simulation. This motivates a data-driven guided search wherein surrogate models trained over data for a small number of oligopeptide chemistries informs an active learning protocol to identify the next most promising chemistries to study. In this work we combine coarse-grained molecular dynamics (CGMD) simulations, variational autoencoders (VAEs), and Gaussian process regression (GPR) to establish an integrated pipeline for the rational exploration of pi-conjugated oligopeptide chemical space. CGMD is used to predict the quality of core-core alignment within the self-assembled nanoaggregates as a measure of fitness, VAEs are trained from scratch to identify a low-dimensional parameterization of chemical space within which to perform the search, and GPR with active learning is used to efficiently navigate the space through an acquisition function balancing exploitation (i.e., chemistries predicted to possess good assembly behavior) and exploration (i.e., searching in under-sampled regions of chemical space). This protocol leads to the identification of new unanticipated pi-conjugated oligopeptide chemistries for all-atom molecular modeling, electronic structure calculations, and experimental synthesis and characterization.

Copyright © American Institute of Chemical Engineers. All rights reserved.

576972 Data-Driven Design of Functional Amyloid Biomaterials

The design of novel amyloid biomaterials with diverse structuresand functions is an upcoming field with future tremendous applications in biomedicine and biotechnology. However, the design of functional amyloid biomaterials with the capacity to bind to ions and compounds, or amyloid biomaterials with cell-attachment properties has been significantly limited until currently experiments primarily on researchers' intuition on which mutations should be added to amyloids to transform them into functional biomaterials. During the last years, we have been developing the first computational protocol with the ability to transform amyloid peptide scaffolds into functional amyloid biomaterials. According to the protocol, suitable amyloid short peptide structures with non-ß-sheet residue positions at the termini (referred to as amyloid designable scaffolds [1]) are transformed into functional amyloid biomaterials, by computationally designing these positions to bear a specific functionality. Specifically, the protocol searches in an optimization-based approach for combinations of modifications, which according to data analysis, can yield a desired functionality and which can produce energetically favorable solutions.

The talk will focus on the design of functional amyloid biomaterials binding to cesium ions, at which the designed amino acids were introduced to the amyloid designable scaffold to mimic how amino acids bind to cesium ions according to data analysis on experimentally resolved structures binding cesium ions [2]. As part of the protocol the optimum designs were computationally validated using a series of simulations and structural analysis to select the top designed peptides which were experimentally validated [2]. The talk will also present recent applications of the computational protocol in additional studies, emphasizing on how data-driven analysis can lead to the formation of functional amyloid biomaterials.

[1] Kokotidou, C.; Jonnalagadda, S. V. R.; Orr, A. A.; Seoane-Blanco, M.; Apostolidou, C. P.; van Raaij, M. J.; Kotzabasaki, M.; Chatzoudis, A.; Jakubowski, J. M.; Mossou, E.; Forsyth, V. T.; Mitchell, E. P.; Bowler, M. W.; Llamas-Saiz, A. L.; Tamamis, P.; Mitraki, A., A novel amyloid designable scaffold and potential inhibitor inspired by GAIIG of amyloid beta and the HIV-1 V3 loop. FEBS Letters 2018, 592 (11), 1777-1788.

[2] Jonnalagadda, S. V. R.; Kokotidou, C.; Orr, A. A.; Fotopoulou, E.; Henderson, K. J.; Choi, C.-H.; Lim, W. T.; Choi, S. J.; Jeong, H.-K.; Mitraki, A.; Tamamis, P., Computational Design of Functional Amyloid Materials with Cesium Binding, Deposition, and Capture Properties. The Journal of Physical Chemistry B 2018, 122 (30), 7555-7568.

Copyright © American Institute of Chemical Engineers. All rights reserved.

573714 Exploration of Interresidual Contacts in Combined Coarse Graining and Experiment Directed Simulations of Aβ42

Experiment Directed Simulations (EDS) and coarse-grained molecular dynamics (MD) were combined to improve the simulation of Aβ42 peptide fragments and to match their computed chemical shifts to experimental results. EDS is a maximum entropy biasing technique that minimally changes (biases) the potential energy of a simulation so that some constraint is satisfied. Constraints are usually an agreement between average position dependent variables from experiments and simulations. EDS has been found to indirectly improve dynamical and structural properties in different systems like peptides, water models, proteins, and lithium ions. EDS has been used in ab-initio, atomistic MD, and coarse-grained MD simulations. Coarse-grained EDS simulations improved the accuracy of actin monomer simulations by reproducing the fluctuations present in the larger actin filaments. In this work, EDS and coarse-graining are used to understand intramolecular and intermolecular interactions in Aβ42. Aβ42 is a 42-residue long amyloid peptide that is associated with Alzheimer's disease. Atomistic MD simulations of multiple chains of 21-30 fragment of amyloid peptide with EDS resulted in their self-assembly previously. Knowing the process of self-assembly in whole Aβ42, including the dynamics and governing interactions will help find therapeutics that halt these processes. In order to study these interactions in Aβ42, we divided Aβ42 into 5 fragments (each fragment consists of 10 residues) with overlapping ends and simulated all of the possible combinations of these 5 fragments in separate simulations. We used chemical shifts from nuclear magnetic resonance spectroscopy results to bias the fragments with chemical shifts from the whole Aβ42 peptide in EDS. All-atom MD simulations were performed with EDS and enhanced sampling in GROMACS simulation engine. We used force matching bottom-up approach to map all-atom forces to a mapped coarse-grained potential. We classified which fragments and which residues interact the most with each other. When EDS bias had been applied, we saw behavior that was predicted in the past experimental literature. During coarse-graining, the interactions were observed between specific fragments of Aβ42 and not with others, which could lead to self-assembly in the full Aβ42.

Copyright © American Institute of Chemical Engineers. All rights reserved.

566350 Conjugation of DOPA and Amyloid Structures for Molecular Design of Underwater Adhesives

Adhesives capable of sticking to various types of surfaces in underwater and high moisture conditions are required for a large array of applications such as marine coatings, sealants, medical devices, and wet living tissue repair. However, typical industrial adhesives are primarily developed for dry applications and perform poorly in wet environments or are highly specific to particular surfaces and not suitable for broad use. A viable and cost-effective strategy to address this issue is to develop generic adhesives that can be used for multiple types of surfaces in water or high moisture conditions.

One approach researchers have taken to develop new adhesives is to draw inspiration from the natural glues produced by aquatic organisms capable of strong, moisture-resistant adhesion to various surfaces. Analysis of these glues has shown that these organisms' abilities to adhere to a multitude of surfaces involve L-3,4-dihydroxyphenylalanine (DOPA) and functional amyloid nanostructures.

The objective of this project is to computationally design bio-inspired underwater adhesives that are stronger than those currently reported in the literature by analyzing and combining the adhesive capabilities of DOPA and amyloid-forming peptides in wet conditions. Explicit solvent atomistic simulations can give insight into the interactions underlying surface adhesion in the presence of water, which can then be used to guide the design of DOPA-amyloid conjugates for underwater adhesion.

Motivated by the concern that DOPA groups might cluster in a way that interferes with their ability to adsorb to surfaces, we investigated the behavior of DOPA monomers in water. In this study, it was found that DOPA was unlikely to cluster significantly at concentrations lower than 1.0 M. DOPA-containing chains (alternating DOPA and glycine groups) were then simulated in water to examine the intra-molecular interactions of the chain, wherein we found that there were unlikely to be interactions detrimental to the adhesion process. We also simulated DOPA molecules on a silica surface and found that they readily adsorb onto silica.

With this understanding of DOPA's behavior under water and at silica surfaces, we began designing a DOPA-amyloid conjugate capable of generic underwater adsorption to varied surfaces. The amyloid-forming peptide KLVFFAE was chosen for the initial design as it is known to self-assemble into functional amyloid nanostructures. Simulations of the conjugated DOPA-amyloid chains showed that the hydroxylated phenyl rings on the DOPA motifs did not interact with the rest of the chain, which leaves them free to bind to the surface. Future work will focus on furthering understanding of the behavior of DOPA-amyloid chains at model surfaces and how changes in the backbone of the DOPA-containing chain and in the sequence of the amyloid-forming peptide affect adhesion on the model surfaces.

Copyright © American Institute of Chemical Engineers. All rights reserved.

563090 Optimizing Antifreeze Protein Activity Via an Elitist Genetic Algorithm: A Comparison of High and Low Activity Variants

Given the existence of a suitably efficient fitness function, genetic algorithms offer a promising method for systematically optimizing the properties of biological materials. In our recent paper [1], we established a neural network based method for predicting the antifreeze activity of different proteins from molecular dynamics simulation. Using this method as our fitness function, we designed a simple, but effective, elitist genetic algorithm for selecting amino acid mutations that should increase the activity of native antifreeze proteins. We apply this algorithm to two antifreeze proteins: (1) a low activity Type III antifreeze protein from the winter flounder, and (2) a hyperactive antifreeze protein from the longhorn beetle. We show that the algorithm finds structures with a predicted freezing point depression 20 - 200% greater than that of the native protein. We hope that these results inspire further investigation of high activity antifreeze biologics that could someday be used in more effective tissue preservation.

[1] Kozuch, D. J., Stillinger, F. H., & Debenedetti, P. G. (2018). Combined molecular dynamics and neural network method for predicting protein antifreeze activity. Proceedings of the National Academy of Sciences, 115(52), 13252–13257

Copyright © American Institute of Chemical Engineers. All rights reserved.

564405 Development of Polymeric Vehicles for Gene Delivery Using High-Throughput Experimentation and Statistical Learning

Complexity is a hallmark of interactions between synthetic materials and the biological environment, but frequently poses barriers to the hypothesis-guided discovery of new materials. Developing polymers for gene delivery is one such multi-dimensional design challenge that requires an intricate balance between intermolecular forces implicated in the packaging, cellular uptake and release of genetic payloads from polymer-nucleic acid complexes. Lacking a complete mechanistic picture of the effects of polymer composition and architecture on its efficiency and toxicity, polymer chemists rely on iterative "one-polymer-at-a-time" approaches that are often plagued by trial and error. Though gene editing has made rapid progress, the paucity of chemically defined, safe and efficient delivery vehicles has hindered its clinical translation. To accelerate polymer discovery for vital clinical applications such as gene editing, we need to replace traditional iterative processes with alternative experimental strategies that leverage advances in parallel synthesis and data analytics. Therefore, we hypothesize that combinatorial polymer libraries, high-throughput experimentation and statistical models can be combined to yield a powerful predictive platform for the discovery and optimization of polymeric vehicles for CRISPR/Cas9 delivery. In this work, a chemically diverse library of well-defined statistical copolymers was synthesized through RAFT polymerization and physicochemical properties such as polymer composition, molecular weight, ς-potential, pK_a, polyplex diameter evaluated. Polymer candidates were screened after complexation with two types of clinically relevant genetic payloads: 1) plasmid DNA and 2) ribonucleoprotein complexes (RNP) comprising a single guide RNA (sgRNA) and Cas9 protein. We suggest that the hit polymers identified through this study could not have been accessible through human intuition and hypothesis-driven approaches. In addition, we probed differences between polymer design criteria for plasmids and RNPs. Using the data obtained through this high-throughput screen, we expect to apply statistical techniques to mine structure-function correlations and predictive relationships that can aid experimentalists engaged in polymer design and synthesis for diverse gene therapy applications.

Copyright © American Institute of Chemical Engineers. All rights reserved.

574920 The Effect of Heteromultivalency on Lectin-Glycan Binding Process and the Identification of Critical Parameter

Glycans are the most diverse fundamental cellular macromolecules that form complex patches over the cell surfaces and bind with lectins (i.e., glycan-binding protein) [1]. The glycan-lectin binding process is important as a wide range of microbial reactions such as bacterial adhesion, viral infection, and cell differentiation, are involved with this [1-3]. One of the unique characteristics of the glycan-lectin binding process is that a single lectin often contains multiple binding pockets interacting with multiple glycan molecules (i.e. multivalency) [1,4]. We recently found that the two dimensional diffusion of glycan molecules on a cell membrane allows a homo-oligomeric lectin simultaneously binds to different types of glycans. This hetero-multivalent binding process alters the lectin binding behaviors, including binding kinetics, energy, and capacity. A living organism may utilize the hetero-multivalency to control the downstream biochemical reactions. In order to understand the role of heteromultivalency in biological processes, a new theoretical approach is needed to model this complex process.

We developed a kinetic Monte Carlo (kMC) simulation, which only relies on the fundamental physics/chemistry principles, to model the process of lectin binding to glycans on cell surfaces. The kMC simulation only considered three universal principles: binding, unbinding and diffusion [5-6]. The kMC simulation allows us to monitor the heteromultivalent binding process and the variation of binding configurations in detail. We observed that the high-affinity glycan ligands can facilitate lectin binding to other low-affinity glycan ligands on a cell membrane. Even though these low-affinity ligands are barely detectable in microarrays where ligands are immobilized, the two dimensional ligand diffusion could activate the interactions between a bound lectin and low-affinity ligands, leading to the changes in binding behaviors. To explore the critical parameters influencing heteromutivalency, we varied the glycan affinity, density, and diffusion rate in the simulations. We observed that a threshold density of low-affinity glycan ligands is required to trigger heteromultivalent binding process. Our work explains why lectin functions sometime are controlled by low-affinity ligands. In contrast to molecular dynamic simulation, the kMC simulation that can model a longer time scales (up to hours) is a better method to study the influence of relatively slow dynamics of glycan diffusion.

References

[1] Varki A. Biological roles of glycans. *Glycobiology*, **2017**, *27*, 3-49.

[2] Branson TR.; McAllister TE.; Garcia-Hartjes J.; Fascione MA.; Ross JF.; Warriner SL.; Wennekes T.; Zuilhof H.; Turnbull WB. A proteinbased pentavalent inhibitor of the cholera toxin B-subunit. *Angew. Chem. Int. Ed. Engl.*, **2014**, *53*, 8323–8327.

[3] Worstell NC.; Singla A.; Saenkham P.; Galbadage T.; Sule P.; Lee D.; Mohr A.; Kwon JSI.; Cirillo JD.; Wu HJ. Hetero-multivalency of Pseudomonas aeruginosa lectin LecA binding to model membranes. *Sci. Rep.*, **2018**, *8*, 8419.

[4] Worstell NC.; Singla A.; Wu HJ. Evaluation of hetero-multivalent lectin binding using a turbidity-based emulsion agglutination assay. *Colloids Surf. B Biointerfaces.*, **2019**, *175*, 84–90.

[5] Fallahi-Sichani M.; Linderman JJ. Lipid raft-mediated regulation of G-protein coupled receptor signaling by ligands which influence receptor dimerization: A computational study. *PLoS One.*, **2009**, *4*, e6604.

[6] Lee D.; Mohr A.; Kwon JSI. Kinetic Monte Carlo modeling of multivariant binding of CTB proteins with GM1 receptor. *Comp. Chem. Eng.*, **2018**, *118*, 283-295.

Copyright © American Institute of Chemical Engineers. All rights reserved.

569751 Big Data from Sparse Data: Diverse Scientific Benchmarks Reveal Optimization Imperatives for Implicit Membrane Energy Functions

In computational protein structure prediction and design, energy functions discriminate non-native from near-native biomolecule conformations. These functions approximate the balance of enthalpic and entropic contributions to protein stability through mathematical models derived from thermodynamic and structural data. Over the past decade, an influx of high-resolution data paired with machine learning techniques has significantly improved the accuracy of soluble protein energy functions. However, membrane protein energy functions remain low-resolution because the experimental data are sparse, leading to overfitting. To overcome this challenge, we assembled a suite of 14 computational benchmark tests against experimental targets. The test probe various membrane protein energy function capabilities ranging from reproducing protein stabilities and orientations to accurately predicting the three-dimensional structures of monomeric proteins and protein complexes. To evaluate current performance, we ran the benchmark on Rosalind19: the current state-of-the-art implicit membrane protein energy function in Rosetta. The benchmark revealed areas for improvement including treatment of electrostatic interactions, representation of the interfacial head group region, and accounting for entropic effects on protein orientation. To enable easy, fast, and continuous of future energy functions, the benchmark suite is publicly available on both GitHub and the Rosetta Benchmark server. Ultimately, these benchmarks will enable big-data-style optimization of membrane energy functions, leading to improved membrane protein design capabilities.

Copyright © American Institute of Chemical Engineers. All rights reserved.

565757 Actively Learning Robust High-Complexity Force Fields

Machine Learning has gained significant traction in the model development community due to its versatility and potential to considerably decrease the human effort required to generate complex, high-fidelity models. Here, we present the Chebyshev Interaction Model for Efficient Simulation (ChIMES), a new machine-learned force field and development approach that has enabled simulation of complex phenomena including condensed-phase reaction-driven phase separation events. ChIMES models are comprised of linear combinations of Chebyshev polynomials explicitly describing many-body interactions and are actively learned to short Kohn-Sham density functional theory (DFT) trajectories. Resulting force fields retain much of the accuracy of DFT while decreasing computational effort by orders of magnitude, and consequently, can be viewed as a proxy for DFT dynamics on large scales.

This work was performed under the auspices of the U.S. Department of Energy by Lawrence Livermore National Laboratory under Contract DE-AC52-07NA27344. LLNL-ABS-771605.

Copyright © American Institute of Chemical Engineers. All rights reserved.

560717 Data Analytics for Designing Fe-9Cr Steels (Tensile Strength)

Motivation for this research comes from the desire to shorten the rigorous and time-consuming alloy qualification (standardization) process, for new fossil energy materials applications. The main consideration for using 9 to 12% Cr martensitic-ferritic steels is their relatively high microstructural stability at the operating temperature over time, since power plants (e.g., steam boiler and turbine applications) have a design lifetime expectation of over 30 years. A materials data analytics (MDA) methodology was developed in this study to evaluate publicly available information on 9% Cr family steel and to handle nonlinear relationships and the sparsity in materials data for this alloy class. Some of the materials research and development activities are proprietary, which makes it particularly difficult to access and compile high-quality information. Data entries in the analyzed data set for over 90 iron-base alloy compositions, several processing parameters, and results of tensile mechanical tests selected for this study were arranged in 47 columns by 2800+ rows. To address non-linearity in the mechanical properties, data analyses were initially carried out in composition-based clusters. A new algorithm was developed for transparent clustering, *c-IG*. The modified *c-IG* using Chebyshev norm produced composition clusters matching the industry classification, without dimensionality reduction. Partitioning revealed the biased nature of available alloy datasets, with implied "rules-of-thumb" in design practices. Processing and test temperature effects on the phases and mechanical properties are the major contributors to predicting the test outcome and need to be modeled first, before searching for minor effects of the composition variations. The evolutionary ensemble models based on the clustering without dimensionality reduction provided a transparent link to the domain knowledge base. While the benchmark linear models (SciPy and TensorFlow) explained less than 50% of the variation in the test data due to controlled alloy design parameters, the ensemble modeling explained over 93% of such variation, for the ultimate tensile strength.

Copyright © American Institute of Chemical Engineers. All rights reserved.

574620 Transfer Learning with Autoencoder Latent Spaces: A Novel Framework for Improved Chemical Prediction from Scarce Datasets

Modern machine learning provides promising methods for accelerating the discovery and characterization of novel chemical species. However, in many areas experimental data remains costly and scarce, and computational models are unavailable for targeted figures of merit. Here we report a promising transfer learning approach to address this challenge by using chemical latent space enrichment, whereby disparate data sources are combined in joint prediction tasks to enable improved prediction in data-scarce applications. The approach is demonstrated for pKa prediction of moderately sized molecular species using a combination of experimentally available pKa data and DFT-based characterizations of the (de)protonation free energy. A novel autoencoder framework is used to create a continuous chemical latent space that is then used in single and joint training tasks for property prediction. By combining these two datasets in a jointly-trained autoencoder framework, we observe mutual improvement in property prediction tasks in the scarce data limit. We also demonstrate an enrichment mechanism that is unique to latent space training, whereby training on excess computational data can mitigate the prediction losses associated with scarce experimental data and advantageously organize the latent space. These results demonstrate that disparate chemical data sources can be advantageously combined in an autoencoder framework with potential general application to data-scarce chemical learning tasks.

Copyright © American Institute of Chemical Engineers. All rights reserved.

572157 Advancing Machine Learning Methodology for the Chemical and Materials Domain

The use of modern machine learning, informatics, and data mining approaches is a relatively new development in the chemical and materials domain. These techniques have been exceedingly successful in other application fields, and since there is no fundamental reason why they should not have a similarly transformative impact on chemical and materials research, there is now a concerted effort by the community to introduce data science in this new context. However, adapting techniques from other application domains for the study of chemical and materials systems requires a substantial rethinking and redevelopment of the existing methods.

In this presentation we will discuss our work on designing advanced, physics-infused neural network architectures, the fusion of unsupervised clustering with supervised regression for local ensemble models, active and transfer learning techniques, bootstrapping approaches to minimize our training data footprint, methods to increase the applicability domain of data-derived models, and automated hyperparameter optimization.

Copyright © American Institute of Chemical Engineers. All rights reserved.

575329 HOOMD-TF: Online Machine Learning in Molecular Dynamics Simulations

We present a novel tool for low-latency online machine learning and collective variable calculation in the HOOMD-blue molecular dynamics (MD) engine. A common issue when pairing machine learning (ML) methods with MD simulations is a high barrier to entry for chemical physicists. This difficulty arises in both offline and online learning situations, in terms of pre- and post-processing of data in the former, or implementation and modification of simulation engine code in the latter. Here, we introduce a solution to this problem that bridges TensorFlow, Google's GPU-accelerated ML library, with HOOMD-blue, a GPU-accelerated MD engine. Our software uses a low-latency GPU-GPU communication scheme to expose the HOOMD neighbor lists of atoms to TensorFlow for reading, and optionally allows TensorFlow to output forces or collective variables calculated from these neighbor lists to HOOMD-blue. This also enables online learning using any value that can be specified as a tensor operation on the HOOMD-blue neighbor list. This gives rise to a wide variety of ML applications, such as coarse-grain force-field learning, calculation of arbitrary collective variables, biasing along arbitrary collective variables, and force-field calculation. In this work, we show four examples of these applications. We train a neural network model to reproduce the Lennard-Jones forces experienced by 10000 Lennard-Jones particles; we train a coarse-grained force field for one-bead methanol on-the-fly during an all-atom simulation; we calculate the neutron scattering profile of 5000 water molecules on-the-fly; and we bias a simulation of Lennard-Jones particles to have a target distance from the center of mass of the system.

Copyright © American Institute of Chemical Engineers. All rights reserved.

564710 Study of Density Functional Errors in Microkinetic Modeling: A Data Driven Approach

Due to their tractable computational cost and relatively high accuracy, Density Functional Theory (DFT) based methods are widely used in computational chemistry to calculate the ground state electronic properties of molecules. Specifically, DFT calculated electronic energies combined with statistical mechanical temperature corrections are used to parametrize microkinetic models of reaction pathways. These *ab-initio* parametrizations involve calculating enthalpies and free energies of molecules in the gas or adsorbed phase. While such models have had some notable successes (for example in identifying dominant reaction pathways), they suffer from highly uncertain rate predictions. A big part of their uncertainty stems from the high sensitivity of rates to uncertainty in *ab-initio* calculated energies. Hence, improving the accuracy of these energies would lead to a substantial improvement in the predictive performance of such hierarchical models. However, it is well known that errors are introduced in DFT calculated properties because of the choice of approximate functionals. These approximations trade accuracy for speed, making the calculation of practical molecules computationally tractable. One way to improve *ab-initio* energy parameters is to introduce systematic corrections to species in the model. These are done to off-set the DFT error. Many such *ad-hoc* methods have been proposed in the literature. However, a systematic framework to introduce such corrections is lacking. In this study, we investigate the errors introduced by commonly used DFT functionals in calculating gas phase properties. Using data-driven techniques we develop a framework to interpret these errors and assign them to easily observable properties of molecules. We thus develop an empirical understanding of systematic errors introduced by DFT calculations in the context of catalytically relevant chemistries. These corrections can easily be applied to specific molecules in the model, improving the overall accuracy of predictions.

Copyright © American Institute of Chemical Engineers. All rights reserved.

571930 Using Molecular Subgraph Libraries for High-Throughput Screening and the Inverse Design Problem

An important set of molecular descriptors in cheminformatics and chemical engineering is the set of molecular subgraphs, sometimes referred to as signature descriptors[1] and used to construct Morgan/circular fingerprints[2]. These molecular subgraphs are of an arbitrary height, K, meaning they contain atomic information about a central atom as well as all atoms within a distance of K bonds from the central atom. In this work, we detail a number of applications of subgraph representations of molecular structures in our computational chemistry software, AMS2019, as well as in the general context of high-throughput screening/inverse design.

First, we discuss the "representability" of large datasets using subgraphs. Specifically, we parse the PubChem database[3] (~100 million compounds) into all possible subgraphs and then calculate the number of subgraphs required to fully represent 100%, 90%, 80%, etc. of the database using subgraphs of different (and mixed) heights. Next, we perform DFT calculations with ADF[4] on the most common 100,000 subgraphs from PubChem and use this library of molecular fragments to generate sigma-profiles for use in high-throughput screening with COSMO-RS[5] as well as in geometry initialization for accelerating DFT calculations. Finally, we discuss the "inverse design" problem: the problem of finding an optimal molecular structure(s) given molecular structure/property constraints and a design objective. We address the inverse design problem using state-of-the-art Mixed-Integer (Non)Linear Programming (MILP/MINLP) techniques[6], exploiting problem structures inherent in subgraph representations of molecules. Specific applications to solvent, drug, and electronic materials design are discussed.

[1] Faulon, Jean-Loup, Donald P. Visco, and Ramdas S. Pophale. "The signature molecular descriptor. 1. Using extended valence sequences in QSAR and QSPR studies." *Journal of chemical information and computer sciences* 43.3 (2003): 707-720.

[2] Rogers, David, and Mathew Hahn. "Extended-connectivity fingerprints." *Journal of chemical information and modeling* 50.5 (2010): 742-754.

[3] PubChem Database. National Institute of Health. https://pubchem.ncbi.nlm.nih.gov/

[4] ADF2018, SCM, Theoretical Chemistry, Vrije Universiteit, Amsterdam, The Netherlands, http://www.scm.com.

[5] Klamt, Andreas, Volker Jonas, Thorsten Bürger, and John CW Lohrenz. "Refinement and parametrization of COSMO-RS." *The Journal of Physical Chemistry A* 102.26 (1998): 5074-5085.

[6] Austin, Nick D., Nikolaos V. Sahinidis, and Daniel W. Trahan. "Computer-aided molecular design: An introduction and review of tools, applications, and solution techniques." *Chemical Engineering Research and Design* 116 (2016): 2-26.

Copyright © American Institute of Chemical Engineers. All rights reserved.

572226 Convolutional Network Analysis of Optical Micrographs for Liquid Crystal-Based Sensors

We use convolutional neural networks to analyze optical responses of liquid crystals (LCs) [1] when exposed to different chemical environments. Our aim is to identify informative features that can be used to construct automated chemical sensors and that can shed some light on the underlying phenomena that govern LC responses. Previous work by Cao and co-workers [2] developed an LC-based chemical sensor that reached accuracy levels of 99% by using spatial and temporal features extracted from the Alexnet convolutional neural network (CNN) [3] and from other basic image analysis techniques such as the histogram of oriented gradients. Unfortunately, reaching such high levels of accuracy required a large number of features (on the order of thousands), which lead to computational issues and clouded the physical interpretability of the dominant features. To address these issues, we study the effectiveness of using features extracted from the VGG16 CNN [4], which is a more compact network than Alexnet. Our findings demonstrate that features extracted from the first and second convolutional block of VGG16 allow for perfect sensor accuracy on the same dataset used by Cao and co-workers while reducing the number of features to less than a hundred. The number of features is further reduced to ten via recursive feature elimination with minimal losses in sensor accuracy. This feature reduction analysis reveals that spatial patterns are developed within seconds in the LC response, which leads to several hypotheses regarding physical mechanisms that underlie sensor selectivity and responsiveness. We also explore the application of dynamic latent variable dimensionality reduction techniques [5] for analysis of the optical micrograph responses.

[1] Shah, Rahul R., and Nicholas L. Abbott. "Principles for measurement of chemical exposure based on recognition-driven anchoring transitions in liquid crystals." *Science* 293.5533 (2001): 1296-1299.

[2] Cao, Yankai, et al. "Machine Learning Algorithms for Liquid Crystal-Based Sensors." *ACS sensors* 3.11 (2018): 2237-2245.

[3] Krizhevsky, Alex, Ilya Sutskever, and Geoffrey E. Hinton. "Imagenet classification with deep convolutional neural networks." *Advances in neural information processing systems*. 2012.

[4] Simonyan, Karen, and Andrew Zisserman. "Very deep convolutional networks for large-scale image recognition." *arXiv preprint arXiv:1409.1556* (2014).

[5] Dong, Yining, and S. Joe Qin. "A novel dynamic PCA algorithm for dynamic data modeling and process monitoring." *Journal of Process Control* 67 (2018): 1-11.

Copyright © American Institute of Chemical Engineers. All rights reserved.

573470 Uncertainty Measurement Method for Machine Learned Potentials

Machine learned potentials (MLPs) have been used to accelerate molecular simulations such as molecular dynamics (MD) and Monte Carlo simulations. Selecting data for training the machine learned models is challenging because the space of atomic structures is often large, not well understood, and not possible to enumerate. In addition, atomic structures are translated into fingerprints which are high dimensional and less human-interpretable than the original atom configurations. Commonly used MLPs such as neural networks will unreliably extrapolate on inputs much different from their training data, and how to select enough of the relevant training space to avoid this is nonobvious. Hence, we require a way to determine the uncertainty of a prediction from a MLP to determine if it is reliable. We propose an uncertainty measurement method for neural network potentials. The uncertainty measure is based on the multiparameter delta method from statistics, which gives the standard error of the prediction. The uncertainty measure is calculated after the MLP is trained, and requires the gradient of the MLP prediction with respect to model parameters and the Hessian of the loss function with respect to model parameters. Both the gradient and Hessian can be obtained from most neural network training frameworks. The method gives a measure of uncertainty for an input that depends on training data, and model parameters. We use examples of intuitive molecular systems to show that the uncertainty measure is large for input space regions not part of the training data. The uncertainty method can be used in combination with MD as an on-the-fly method, and it can aid in selecting training data.

Copyright © American Institute of Chemical Engineers. All rights reserved.

575637 An Automated Job Management Framework for High Fidelity Quantum Chemistry Calculations

Recent advancements in quantum chemistry methods allow chemical properties to be calculated *ab-initio* within chemical accuracy (1 kcal/mol). With the advance of High Performance Computing the bottleneck is no longer the computation, but the human interaction required to set up calculations, curate, and interpret the results. Calculating thermochemical kinetics accurately requires multiple interdependent calculations, for example to search for conformers, optimize geometry, calculate electronic correlation, calculate valence and core electrons separately, perform hindered rotor scans, etc. Automation of these calculations with minimal human interaction is needed to realize the full potential of ever-increasing computational power in many research fields. Several workflows and protocols have been proposed to automate these calculations, but not for actual job spawning, managing dependencies between these jobs, and post processing results automatically.

In this talk, we present an automated framework that automatically spawns jobs, and handles dependencies between jobs, on multiple supercomputing clusters, with minimal user interaction. Specifically, we use FireWorks [1], an open source automated framework written in Python for defining, managing, and executing complex workflows. Jobs along with their dependencies are stored in MongoDB. FireWorks consists of two components, a FireServer that manages workflows and a FireWorker that executes these workflows. We demonstrate the efficacy of FireWorks through automated scheduling of high-fidelity quantum chemistry calculations for the construction of detailed kinetic models. This will enable accurate thermodynamic and kinetic parameters to be calculated in a high-throughput manner that makes the most of computational resources available to a user. In addition, this framework allows for the creation of a standardized dataset that can be used as training data for machine-learning based estimation methods.

References

1. Jain, A., Ong, S. P., Chen, W., Medasani, B., Qu, X., Kocher, M., Brafman, M., Petretto, G., Rignanese, G.-M., Hautier, G., Gunter, D., and Persson, K. A. (2015) FireWorks: a dynamic workflow system designed for high-throughput applications. Concurrency Computat.: Pract. Exper., 27: 5037–5059. doi: 10.1002/cpe.3505.

Copyright © American Institute of Chemical Engineers. All rights reserved.

562353 Theoretical Investigation of Pt-Catalyzed Dehydrogenation of Propane to Propylene: Applying Uncertainty Analysis to Reaction Kinetics

Results of density functional theory (DFT) based-models for reaction systems are functional dependent. Changing the type of functional used for the model can often change the predictions of the model, such as the turnover frequency, reaction orders, apparent activation energy, and ultimately what active site dominates for the reaction in question. Previous work on creating a framework and methodology for quantifying uncertainty has been shown to help elucidate how certain reactions occur, and for generating a scheme for comparing different models in order to identify the most likely active site for reaction[1,2].

The reaction kinetics of non-oxidative propane dehydrogenation (PDH) of propane to propylene was investigated over a series of 3 surface models, Pt(100), Pt (111), and Pt(211), using plane wave density functional theory calculations. In these calculations, we used four different functionals, which we then carried out different statistical analyses to solve a statistical forward problem (SFP) using the Quantification of Uncertainty for Estimation, Simulation and Optimization program (QUESO), in order to generate the PDH reaction's quantities of interest, including the turnover frequency, reaction orders, and the apparent activation energy, at low conversion and low coking reaction conditions with a temperature of 633 K, based on previous experimental work[3]. To identify to impact of functional selection in the prior generation, we also use the energy ensembles generated by the Bayesian Error Estimation Functional with van der Waal corrections (BEEF-vdw) to generate another uncertainty model[4]. We created a lateral interaction model and a full hydrogen coverage model in order to replicate the reaction conditions, and generate data that may support that Pt(100) might be the active surface that non-oxdiative propane dehydrogenation occurs on in these conditions.

References

[1] *J. Phys. Chem. C*, **2016**, *120* (19), pp 10328–10339

[2] *ACS Catal.*, **2018**, *8* (5), pp 3990–3998

[3] *J Catal*. **1977**, 50, pp 77-88

[4] *Phys. Rev. B*, **2012** 85, 235149

Copyright © American Institute of Chemical Engineers. All rights reserved.

563360 Symbolic Regression for the Automated Physical Model Identification in Reaction Engineering

Symbolic regression for the automated physical model identification in reaction engineering

Liwei Cao[a,b], Pascal Neumann[a,c], Danilo Russo[a], Vassilios S. Vassiliadis[a], Alexei A. Lapkin [a,b*]

[a] Department of Chemical Engineering and Biotechnology, University of Cambridge, Philippa Fawcett Drive, Cambridge CB3 0AS, U.K.

[b] Cambridge Centre for Advanced Research and Education in Singapore (CARES) Ltd., Campus for Research Excellence and Technological Enterprise (CREATE), CREATE Tower, 1 CREATE Way, Singapore, 138602

[c] Aachener Verfahrenstechnik - Process Systems Engineering, RWTH Aachen University, Aachen 52062, Germany

* **Corresponding author; email:** aal@cam.ac.uk

Abstract

Understanding of a complex reaction system at a fundamental level is crucial as it reduces the time and resources required for process development and implementation at scale. The two distinct paradigms in developing fundamental knowledge of a chemical system start from either experimental observations (data-driven modeling), or from mechanistic *a priori* knowledge (physical models). With the rise of automation and tremendous modern advancements in data science the two approaches are gradually merging, although model identification for multivariable complex systems remains challenging in practice. In this work, the identification of interpretable and generalizable physical models is targeted by means of automatable, data-driven methods without *a priori* knowledge. A revised mixed-integer nonlinear programming (MINLP) formulation is proposed for symbolic regression (SR) to identify physical models from noisy experimental data. The identification of interpretable and generalizable models was enabled by assessing model complexity and extrapolation capability. The method is demonstrated by successful application for the identification of a kinetic model of the 4-nitrophenyl acetate (PNPA) hydrolysis reaction.

Copyright © American Institute of Chemical Engineers. All rights reserved.

563971 A Machine Learning Based Algorithm for Rate Estimation

A Machine Learning Based Algorithm for Rate Estimation

Matthew S. Johnson[1] and William H. Green[1,*]

[1]Dept. Chemical Engineering, Massachusetts Institute of Technology,

Cambridge, MA-02139 – USA

Large chemical kinetic mechanisms are vital to understanding many complex chemical processes but can involve the estimation of >100,000's of reaction rates, far too many to determine from experiments or quantum chemical methods. Modern ab initio methods can now provide us with enough kinetic data to approach this problem from a machine learning perspective. However, rate estimation is highly nonlinear and kinetic datasets are very sparse. For these reasons we need a demanding combination of properties which have previously been unattainable from conventional machine learning approaches: nonlinear, easy to diagnose, and capable of incorporating qualitative chemical knowledge and uncertainty estimation. We have developed a subgraph-isomorphic decision-tree based method that has all of these properties. The only inputs for construction are a set of kinetic data for training and a starting reaction template. Each node of the tree contains the set of training reactions matching the corresponding reaction template. Their rate coefficients, k, are treated as samples from a lognormal rate distribution. Each node is branched into sub-nodes using more specific reaction templates chosen to minimize the sum of the standard deviations in log(k). Subgraph properties that do not affect this sum (usually because the training reactions are too homogeneous) are identified and adjusted based on a chemical knowledge based algorithm. Rates and uncertainties can then be estimated treating each reaction rate as a sample of the distributions at each node. A comparison was run against data and estimators in the Reaction Mechanism Generator (RMG) database. Using this technique, training on the RMG-database training sets, decision trees were generated for the Subsitution_O (127 reactions) and intra_H_migration (422 reactions) reaction types. Comparing against RMG's rate rules at T=1000 K the decision tree estimator reduced 2-sigma errors by a factor of 18 and 45 respectively. Furthermore, median errors lumping all error into the activation energy were reduced from 5.1 to 3.1 kcal/mol and from 4.0 to 2.9 kcal/mol respectively.

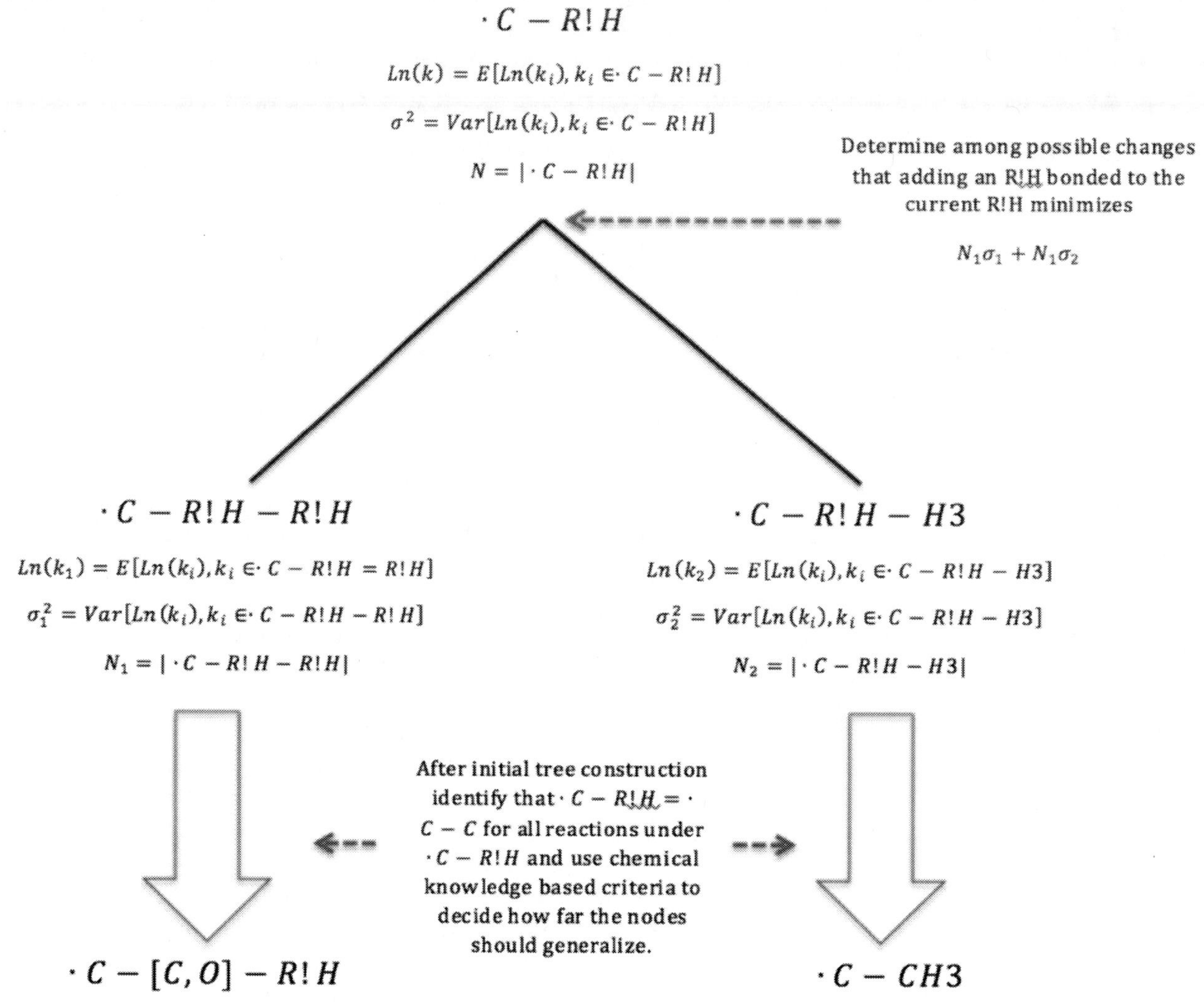
·C − R!H
Ln(k) = E[Ln(k_i), k_i ∈· C − R!H]
σ^2 = Var[Ln(k_i), k_i ∈· C − R!H]
N = |·C − R!H|
Determine among possible changes that adding an R!H bonded to the current R!H minimizes
N_1σ_1 + N_1σ_2
·C − R!H − R!H
Ln(k_1) = E[Ln(k_i), k_i ∈· C − R!H = R!H]
σ_1^2 = Var[Ln(k_i), k_i ∈· C − R!H − R!H]
N_1 = |·C − R!H − R!H|
·C − R!H − H3
Ln(k_2) = E[Ln(k_i), k_i ∈· C − R!H − H3]
σ_2^2 = Var[Ln(k_i), k_i ∈· C − R!H − H3]
N_2 = |·C − R!H − H3|
After initial tree construction identify that ·C − R!H = ·C − C for all reactions under ·C − R!H and use chemical knowledge based criteria to decide how far the nodes should generalize.
·C − [C, O] − R!H
·C − CH3

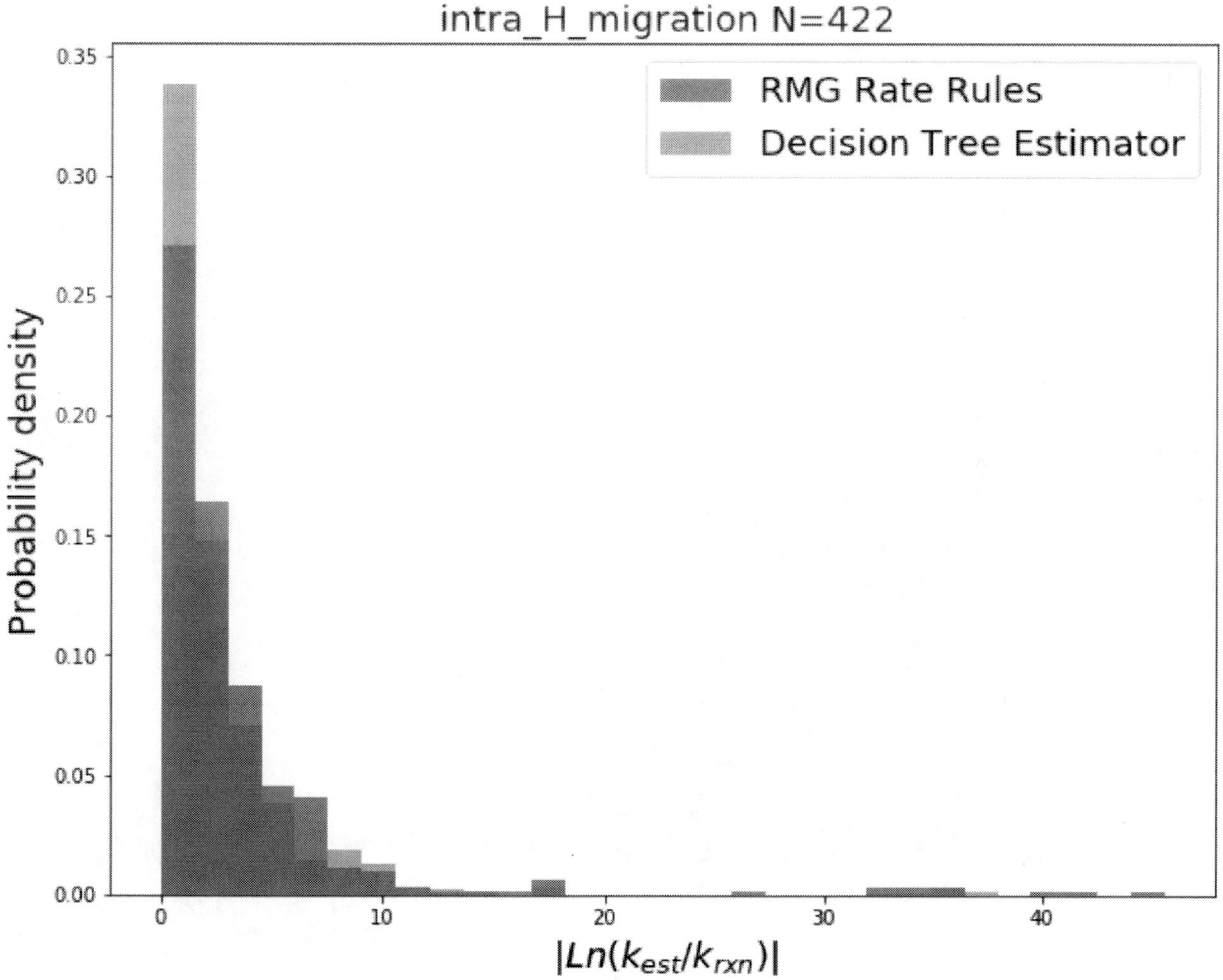
intra_H_migration N=422
RMG Rate Rules
Decision Tree Estimator
Probability density
0.35
0.30
0.25
0.20
0.15
0.10
0.05
0.00
0
10
20
30
40
|Ln(k_est/k_rxn)|

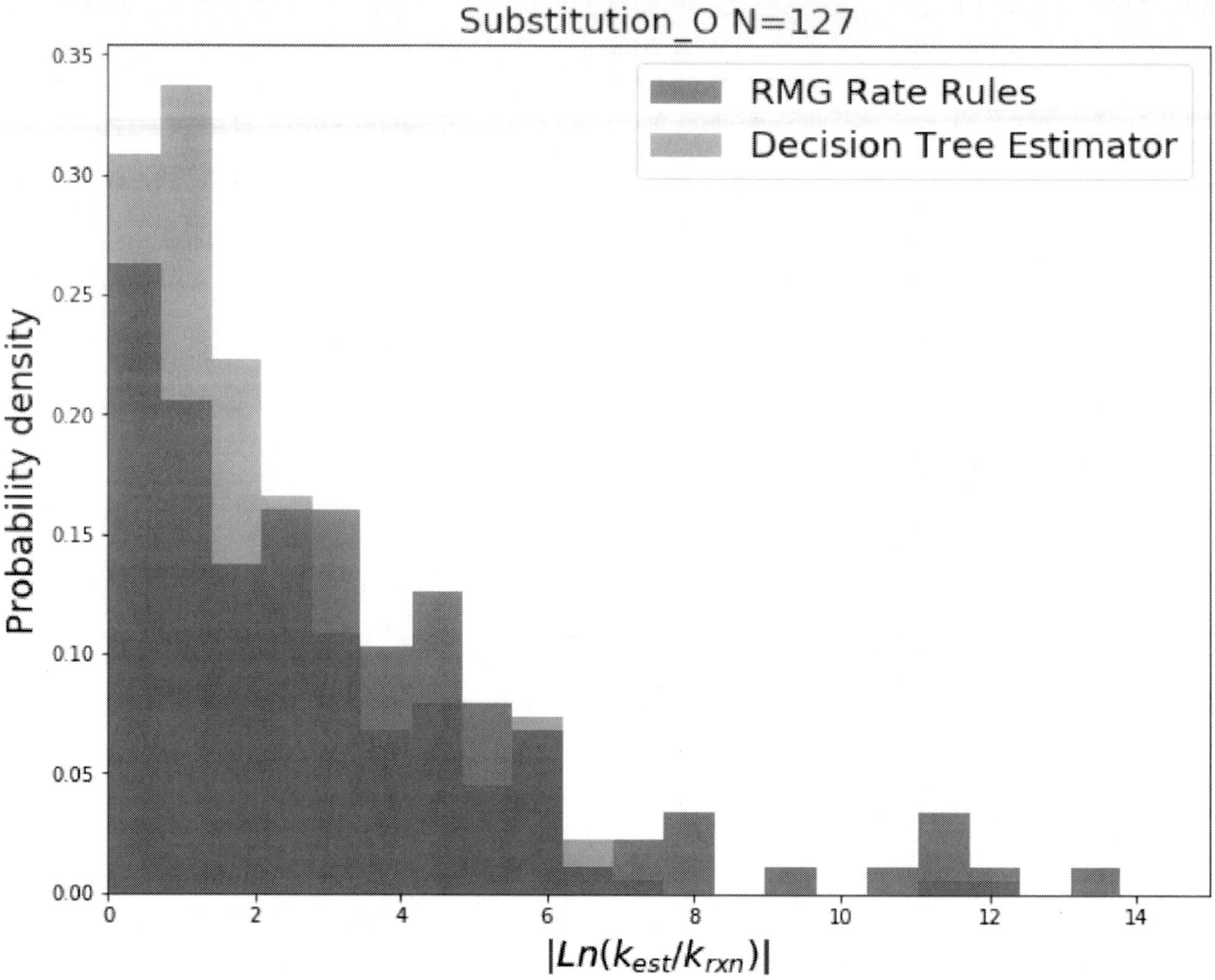

Copyright © American Institute of Chemical Engineers. All rights reserved.

565904 Using Data Driven Models to Gain Insight on Spin- and Oxidation-State Dependent Behavior of Reaction Energetics for Light Alkane Oxidation

Computational high-throughput virtual screening (HTVS) with first-principles density functional theory (DFT) can play a valuable role in unearthing design rules for scalable, industrially viable synthetic analogues that preserve selectivity and activity seen only in enzymes. Single-site catalysts represent the most promising analogues to these enzymes, often enabling atom-economy and selectivity not possible with bulk heterogeneous catalysts. Simultaneously, a wide chemical space must be explored in the realm of single-site catalysts, even beyond the reach of conventional DFT screening, in order to simultaneously address other constraints, such as high turnover number or robustness, earth abundance, and synthesizability. Single-site catalysts have the added dimensionality of spin- and oxidation-state, which can drastically impact the structure-property relationships of molecular complexes and remain unexplored for catalyst reaction energetics. We demonstrate our developments on data-driven models for key steps in single-site light alkane oxidation catalysts, which enable the prediction of reaction energetics in a spin- and oxidation-state dependent manner close to the accuracy of DFT. We first compare prediction of reaction energetics to other quantum mechanical properties that we have predicted, such as spin-splitting energetics, ionization potential, and frontier orbital energetics. We then discuss insights on representative catalytic reaction steps such as oxo formation and hydrogen atom transfer, including their dependence on spin- and oxidation-state. Lastly, we demonstrate the power of data-driven models for purposes of screening, enabling screens of large design spaces that would be infeasible to screen by DFT, even with the reduction in number of calculations through the use of descriptor energies. Having separate data driven models for different reaction energy steps allows predictions of reaction energies that contain weak or nonexistent linear free energy relationships (LFERs), removing dependence on LFERs that have been shown to be broken in single-site catalysis.

Copyright © American Institute of Chemical Engineers. All rights reserved.

566128 Statistical-Learning-Assisted First-Principles Modeling of Catalysts Sintering

Supported metal nanoparticles on oxide supports are widely applied in numerous applications. The catalyst activity is often strongly affected by the nanoparticle size. However, predicting catalyst sintering has remained elusive in part due to the inability to observe the dynamics of the catalysts operando and in part because adsorbates affect catalyst dynamics and sintering [1]. Density functional theory (DFT) is a powerful tool to predict the energetics of metal-support and adsorbate-metal interactions. However, the computational time needed to describe the long-time scales for sintering makes direct first-principles calculations impractical. To address this multiscale problem, here we introduce statistical learning tools to rapidly predict the parameters of Pd sintering supported on $CeO_2(111)$ using input from DFT calculations. Kinetic Monte Carlo (kMC) simulations are applied to model the catalyst sintering. The methodology is applicable to any metal/support system.

[1] Wang, Y., Mei, D., Glezakou, V., Li, J. & Rousseau, R. Nat. Commun. 6, 1–8 (2015).

Copyright © American Institute of Chemical Engineers. All rights reserved.

566423 Data-Driven Strategy for Ab-Initio Microkinetic Modeling of C1-Reaction on Copper (111) from Disparate Experimental Datasets

Microkinetic models based on energies derived from density functional theory (DFT) (with or without re-parameterization with kinetic data) are often used to elucidate mechanisms and quantify reaction fluxes of catalytic systems. However, such analyses are often carried out for a specific system of interest and independent of related chemistries.

Here, we present a statistical framework that enables a systematic and active improvement of a microkinetic model by making use of a variety of thermokinetic data (e.g. adsorption calorimetry, DFT energies, steady-state kinetics, and temperature-programmed studies). To achieve this, we build a statistical framework that systematically calibrate the intrinsic error of DFT and quantify the uncertainty. Then, Bayesian experimental design is used to improve the prediction of kinetic model by sequentially select the most "informative" kinetic experiments of different type and from different resources.

We demonstrate how this approach allows us to build a unified model describing the reactions of C1 oxygenates on Cu(111). This work demonstrates the feasibility (and caveats) of kinetic modeling of large reaction network using disparate datasets.

Copyright © American Institute of Chemical Engineers. All rights reserved.

576465 Generalized Brønsted Evans Polanyi Relationships for Metal Surfaces from Machine Learning

In state of the art catalyst design approaches the largest cost is associated with the parameterization of Brønsted-Evans-Polanyi (BEP) relationships [1], a method to correlate kinetic parameters with thermodynamic ones, at the elementary step level. Therefore, the development of a generalized formulation of BEP correlations, i.e. BEPs that can accurately describe a wide array of reactions, would be highly desirable and would accelerate catalyst design.

Here we develop generalized BEPs for a diverse set of reactions on metal surfaces. We focus on 788 reactions from the CatApp database [2], a database that contains reaction- and activation energies for reactions on monometallic surfaces calculated using the DACAPO code and using the RPBE density functional. Our analysis builds on typical BEPs, which only use the reaction energy as a descriptor. However, to improve accuracy we introduce a set of descriptors for the type of reaction encountered derived from scaling relationships [3] and add an additional descriptor for the catalyst surface. In a second step we use three different machine-learning approaches for inter- and extrapolation. We find a significant increase in accuracy for the prediction of activation energies and our best approach leads to errors that are comparable to errors typically associated with DFT calculations. In further analysis, we find that MAE values are similar for different types of reactions and we study the impact of training set size on the performance of the approach. Finally we confirm that the reaction energy is the most important descriptor, but all the other introduced parameters collectively contribute significantly to the improved description.

[1] J.K. Nørskov, T. Bligaard, A. Logadotir, S. Bahn, L.B. Hansen, M. Bollinger, H. Bengaard, B. Hammer, Z. Sljivancanin, M. Mavrikakis, Y. Xu, S. Dahl, C.J.H. Jacobsen, *J. Catal.* **209**, 275-278 (2002)

[2] J.S. Hummelshøj, F. Abild-Pedersen, F. Studt, T. Bligaard, J.K. Nørskov, *Angew. Chemie - Int. Ed.* **51,** 272–274 (2012).

[3] F. Abild-Pedersen, J. Greeley, F. Studt, J. Rossmeisl, T.R. Munter, P.G. Moses, E. Skulason, T. Bligaard, J.K. Norskov, Phys. Rev. Lett. 99, 016105, 2007

Copyright © American Institute of Chemical Engineers. All rights reserved.

561273 Understanding Photonic Band Gaps in Three Dimensions

Photonic crystals are materials composed of mixed dielectric media that result in the reflection of all electromagnetic waves within a range of frequencies (called a complete photonic band gap or "stop-band") commensurate with the length scale of the crystal. This scientific advance poses an exciting avenue for novel materials---they allow for light to be controlled through materials design, and their implementation in technology has been likened to the introduction of semiconductors. Since first theorized in 1987, there has been a rush to define and synthesize photonic crystal structures. In the decades since, many photonic structures have been discovered, often by using naturally occurring crystal structures as templates for design. However, these studies have yet to answer the question: what features of a 3D structure will produce a photonic band gap? Here, we compile a large database of periodic structures templated from a variety of naturally occurring crystal and calculate their photonic band structure using MIT Photonic Bands (MPB). Our simulations suggest that when designing novel photonic materials, the toolbox of structural templates may be larger and richer than previously thought, widening the field of target crystal structures.

Copyright © American Institute of Chemical Engineers. All rights reserved.

577441 Learning Structure Identification in Molecular Simulations with a Pointnet

Molecular simulations are applied to study increasingly diverse phenomena in physics, chemistry, biology, and engineering. The raw simulation output (i.e., positions of each particle in the system) contains a wealth of data about the system which can be used to relate microscopic structure to macroscopic observable properties and processes. As such, identifying local structure is a key feature of the analysis of molecular simulations. The most common existing approach to identify local structure is to calculate some geometrical quantity referred to as an order parameter. In some simple cases order parameters are physically intuitive and trivial to develop (e.g., ion-pair distance), however in most cases, order parameter development becomes a much more difficult endeavor (e.g., collective solvent behavior around ions). The difficulty of order parameter development is evidenced by the observation that a new order parameter is often worthy of a publication.

Given the widespread success of machine learning in object identification in fields such as computer vision, it seems intuitive that similar approaches might hold promise for structure identification in simulations. In general, the challenge in applying machine learning techniques to simulation is selecting the appropriate input features. This once again becomes a system-specific challenge that can require significant human input and intuition. Our goal is to develop a generic approach to local structure identification in molecular simulations that requires no system-specific feature engineering and operates on the raw outputs of the simulations, i.e., atomic positions.

Our approach is to apply a specific type of neural network called a PointNet to identify local structural environments in molecular simulations. The PointNet takes as input the relative positions (i.e., x, y, z) coordinates of atoms within some cutoff distance of a central atom and is trained to classify the structural environment of the central atom. We demonstrate the method on crystal structure identification in Lennard—Jones (four different phases), water (eight different phases), and mesophase (six different phases) systems. The method is able to achieve >99.5% accuracy in crystal structure identification. We also demonstrate that the method is applicable to heterogeneous nucleation as it can even predict the crystal phases of atoms near external interfaces. Next, we test if the method can be used to identify if a water molecule is in a hydrophobic or hydrophilic environment based solely upon the surrounding water molecules. The method is able to characterize heterogeneous surfaces (e.g., self-assembled monolayer surfaces with patterns of hydrophobic and hydrophilic regions) and can characterize the local surface hydrophobicity of biomolecules such as proteins.

The PointNet approach to local structure identification is a generic framework for identifying different structural motifs which appear in molecular simulations. The approach only requires the raw output from molecular simulations and requires no system-specific feature engineering. The approach will be broadly applicable to many other types of local structure in simulations, such as biomolecular conformations, active site arrangements, the structure of solvation shells, and more. The method will shorten time-to-discovery by enabling rapid structure identification in novel systems where there are not existing order parameters.

Copyright © American Institute of Chemical Engineers. All rights reserved.

576341 Studying the Importance of Preserving Symmetry in Coarse-Grained Alkanes

In this study we compare different symmetric and asymmetric coarse-grained (CG) mapping operators to understand the importance of preserving symmetry while choosing a CG mapping operator. CG simulations have been widely used to study systems that span large spatiotemporal scales. Selecting a CG mapping operator and obtaining the corresponding potentials are the key steps towards defining a CG model. Both these choices significantly affect how closely a CG simulation reproduces results from the corresponding all-atom (AA) simulation. While there has been considerable progress towards systematically deriving CG potentials, the choice of CG mapping operator is conventionally made using chemical intuition. Recently, there have been efforts towards developing more systematic approaches for defining a CG mapping operator. There is a lack of comprehensive studies on different factors that might influence the efficiency of CG mapping operators. Here we compared different CG mapping operators of alkanes and their isomers to investigate if it is important to preserve the symmetry present in the original alkane molecule while choosing a CG mapping operator. We considered two alkanes, hexane and octane, and three of their isomers. For each isomer, symmetric and asymmetric CG mapping operators were used to perform bottom-up CG simulations. The CG mapping operators were evaluated based on how closely the CG simulation results matched with the structural and dynamic properties obtained from the corresponding all atom simulation. We explored unsupervised cluster analysis as an evaluation metric for CG simulations. This study would contribute towards developing rigorous and systematic strategies, going beyond chemical intuition, to choose effective CG mapping for a molecule.

Copyright © American Institute of Chemical Engineers. All rights reserved.

576821 Construction of Coarse-Grained Neural Network Potentials for Predicting Water Properties

Accurate yet computationally trackable interatomic potential will be invaluable in many fields of science and engineering. However, many existing interatomic potentials are frequently limited by their fixed functional form, high computational cost, transferability for a wide range of environmental conditions. In particular, the transferability of force field is very poor for polar and hydrogen bonded liquids such as water. Recent advances in machine learning methods have shown promising pathways to improve the accuracy and transferability of inter-atomic potentials. In this direction, here, we attempt to develop a coarse-grained neural network (CG-NN) potential that predict the potential energy surface (PES) of water very accurately. Although the atomic forces are not trained to the network, the analytical derived forces from the predicted energies shows appreciable agreement with the actual atomic forces on the coarse-grained atoms. Further Monte Carlo (MC) and molecular dynamics (MD) simulations based on CG-NN potential predict reasonably accurate thermophysical properties of water. We also analyze the correlations between the predictability of the CG-NN model with its topology and the choice of symmetry functions that represent a PES.

Copyright © American Institute of Chemical Engineers. All rights reserved.

560202 A Machine Learning Guided Approach for Studying Local Solvation Environments

In this work we introduce a calculation scheme that uses state of the art computational tools to study ion solvation environments. We first generate microsolvated clusters with global optimization algorithm called ABCluster and then introduce a measure of convergence based on the structural similarities of solvent molecules near the solute using unsupervised machine learning algorithms called Smooth Overlap of Atomic Positions (SOAP) and sketchmaps. SOAP is a clustering technique that identifies structures that are closely related to each other, and the sketchmaps reduce the dimensionality of the feature vectors to represent the SOAP kernels. After studying different solvation environments, we calculate solvation energies using Quasi Chemical Theory (QCT) and thermodynamic cycles. We demonstrate that low energy molecular clusters produced by our procedure have structurally similar local solvation environments, and this suggests that this calculation scheme can be used to quantify the number of explicit solvent molecules needed to accurately model the relevant local solvation environment of a charged solute. We expect this approach will be broadly applicable for the atomistic modeling of ions in different solvents.

Copyright © American Institute of Chemical Engineers. All rights reserved.

564389 Toward Universal Prediction of Adsorption Isotherms in Metal-Organic Frameworks Using Deep Learning Algorithms

Machine learning is emerging as a tool that can further accelerate the screening of material databases over traditional molecular simulation. For example, machine learning models can be trained to predict the adsorption properties of metal-organic frameworks (MOFs) for applications such as gas storage and chemical separations. However, previous efforts to predict adsorption using machine learning have focused on developing algorithms using simulated data for a specific application at a single operating condition. Therefore, if the need emerges to evaluate materials for a different operating condition (e.g. different pressure, temperature or gas composition), or for a different application (involving different adsorbates) extensive molecular simulation needs to be done again to enable the training of a new algorithm. To overcome this issue, in this contribution we demonstrate the viability of predicting adsorption loadings for different adsorbates at different conditions using a single model.

To do so, we demonstrate the prediction of full isotherms (in the 1-100 bar range) of small adsorbate gases in a sample MOF database. These predictions were made possible by introduction of adsorbate properties as part of the training data (in addition to MOF properties). Notably, this approach allows our model to predict the loading of adsorbates not included in the training set. We trained our model on a database of 2,000 topologically and chemically diverse MOFs using simulated data for nine different atomic and diatomic adsorbates. Models were trained by a "leave two out" approach. Briefly, in this approach, performance was evaluated by using a version of the model trained on only eight of the adsorbates and a subset of the MOFs to predict the loading of the "left-out" adsorbate in the "left-out" MOFs. That is, no adsorbate or MOF used to train each model version was included in the data used to validate that version. We will discuss the descriptor set (among five tested) that yields the most accurate loading predictions.

Among "as is" algorithms, we found multilayer perceptrons (MLPs) to outperform other models. However, to accomplish the desired accuracy, we found that a hierarchical "committee of nets" approach was appropriate. Briefly, the "committee of nets" consists of a top-level MLP that makes loading predictions using as input the predictions of multiple individual base-level MLPs along with the descriptors used to train these base models. Given its accuracy and scope, our final model can be coupled with traditional mixture-adsorption prediction approaches to rapidly estimate (orders of magnitude faster than traditional molecular simulations) the adsorption loadings of diverse gas mixtures relevant to diverse chemical separations in MOF databases. The combination of these capabilities with our Topologically Based Crystal Constructor (ToBaCCo) code for computational synthesis of MOF prototypes could allow high throughput screening of MOFs at previously inaccessible scales.

Copyright © American Institute of Chemical Engineers. All rights reserved.

568992 Intensive Chemical Descriptors for Nanoporous Crystalline Materials

Molecular descriptors play a key role in the development of predictive structure-property/activity relationships (QSP/ARs) that aid molecular discovery/design. Although thousands of molecular descriptors have so far been proposed, the range of descriptors that have been explored for nanoporous crystalline materials such as metal-organic frameworks (MOFs) has been limited to few mostly macroscopic geometrical descriptors. In this study, we introduce a new robust procedure to generalize the thousands of descriptors that already known from molecular systems to crystalline nanoporous materials. We will show how these descriptors have enabled machine-learning approaches to efficiently and accurately predict adsorption isotherms for diverse collections of thousands of molecules in libraries of thousands of MOFs.

Copyright © American Institute of Chemical Engineers. All rights reserved.

574579 Eigencages: Learning a Latent Space of Porous Cage Molecules

Porous organic cage molecules harbor nanosized cavities that can selectively adsorb gas molecules, lending them applications in separations and sensing. The geometry of the cavity strongly influences their adsorptive selectivity. For comparing cages and predicting their adsorption properties, we embed/encode a set of 74 porous organic cage molecules into a low-dimensional, latent "cage space" on the basis of their intrinsic porosity. We first computationally scan each cage to generate a three-dimensional (3D) image of its porosity. Leveraging the singular value decomposition, in an unsupervised manner, we then learn across all cages an approximate, lower-dimensional subspace in which the 3D porosity images congregate. The "eigencages" are the set of orthogonal, characteristic 3D porosity images that span this lower-dimensional subspace, ordered in terms of importance. A latent representation/encoding of each cage follows by approximately expressing it as a combination of the eigencages. We show that the learned encoding captures salient features of the cavities of porous cages and is predictive of properties of the cages that arise from cavity shape. Our methods could be applied to learn latent representations of cavities within other classes of porous materials and of shapes of molecules in general.

Copyright © American Institute of Chemical Engineers. All rights reserved.

568020 Deep Neural Network Learning of Complex Binary Sorption Equilibria from Molecular Simulation Data

Deep neural networks (NNs) is employed as an efficient and intelligent surrogate of molecular simulations for complex sorption equilibria using probabilistic modeling. Canonical (N_1N_2VT) Gibbs ensemble Monte Carlo simulations were performed to model a single-stage equilibrium desorptive drying process for (1,4-butanediol or 1,5-pentanediol)/water and 1,5-pentanediol/ethanol from all-silica MFI zeolite and 1,5-pentanediol/water from all-silica LTA zeolite. A multi-task deep NN was trained on the simulation data to predict equilibrium loadings as a function of thermodynamic state variables. The NN accurately reproduces simulation results and is able to obtain a continuous isotherm function. Its predictions can be therefore utilized to facilitate optimization of desorption conditions, which requires a laborious iterative search if undertaken by simulation alone. Furthermore, it learns information about the binary sorption equilibria as hidden layer representations. This allows for application of transfer learning with limited data by fine-tuning a pretrained NN for a different alkanediol/solvent/zeolite system.

Copyright © American Institute of Chemical Engineers. All rights reserved.

AUTHOR INDEX

AUTHOR INDEX